# Springer Tracts in Additive Manufacturing

**Series Editor**

Henrique de Amorim Almeida, Polytechnic Institute of Leiria, Leiria, Portugal

D1826908

The book series aims to recognise the innovative nature of additive manufacturing and all its related processes and materials and applications to present current and future developments. The book series will cover a wide scope, comprising new technologies, processes, methods, materials, hardware and software systems, and applications within the field of additive manufacturing and related topics ranging from data processing (design tools, data formats, numerical simulations), materials and multi-materials, new processes or combination of processes, new testing methods for AM parts, process monitoring, standardization, combination of digital and physical fabrication technologies and direct digital fabrication.

More information about this series at http://www.springer.com/series/16694

Kamalpreet Sandhu · Sunpreet Singh ·
Chander Prakash · Karupppasamy Subburaj ·
Seeram Ramakrishna
Editors

# Sustainability for 3D Printing

 Springer

*Editors*
Kamalpreet Sandhu
Department of Product and Industrial Design
Lovely Professional University
Phagwara, Punjab, India

Chander Prakash
Department of Industrial Engineering
Lovely Professional University
Phagwara, Punjab, India

Seeram Ramakrishna ⓘ
Department of Mechanical Engineering
National University of Singapore
Singapore, Singapore

Sunpreet Singh ⓘ
National University of Singapore
Singapore, Singapore

Karupppasamy Subburaj
Engineering Product Development
Singapore University of Technology
and Design
Singapore, Singapore

ISSN 2730-9576          ISSN 2730-9584  (electronic)
Springer Tracts in Additive Manufacturing
ISBN 978-3-030-75237-8          ISBN 978-3-030-75235-4  (eBook)
https://doi.org/10.1007/978-3-030-75235-4

This Springer imprint is published by the registered company Springer Nature Switzerland AG
The registered company address is: Gewerbestrasse 11, 6330 Cham, Switzerland

# Preface

The book entitled *"Sustainability for 3D Printing"* aimed to present various experimental outbreaks on the novel methodologies to treat solid waste as different types of feedstock materials, suitable for demanding design and engineering applications, of 3D printing technologies. This book provides a comprehensive knowledge of the innovative models, machine tools, and processing routes adopted for treating solid wastes and recycling/reuse of the same as different types of 3D printing feedstock. In particular, detailed discussions on the life cycle assessment, sustainability, and eco-friendliness of the developed feedstock as well as end user products also be incorporated. The editorial team has understood that number of books already published in the field of the 3D printing with different focuses; however, it is of utmost important to produce an ideal and reader friendly source of literature with "Waste to Wealth" focus as it have huge potential to serve as a reference source to the experts of manufacturing, materials, metallurgy, product design, waste management, and sustainability prospective. Further, it has been believed by the editorial members that every manufacturer, in today's manufacturing era, is bounded to follow the sustainability ethics to minimize the negative impact of manufacturing on environment as well as well-being of the living species. Indeed, this edited book provides wide variety of literature review, case studies, experiential studies, and technical papers to highlight the scope of using waste as wealth through 3D printing.

Phagwara, India — Kamalpreet Sandhu
Singapore, Singapore — Sunpreet Singh
Phagwara, India — Chander Prakash
Singapore, Singapore — Karupppasamy Subburaj
Singapore, Singapore — Seeram Ramakrishna

# Contents

# About the Editors

**Kamalpreet Sandhu** is an Assistant Professor in the Product and Industrial Design department at Lovely Professional University, Phagwara, Punjab, INDIA. His primary focus is on design and developed of footwear products and injuries prevention. He was done various projects in Podiatric Medicine at Defence Institute of Physiology and Allied Sciences, DRDO, Delhi, i.e. design and developed new kind of orthosis for social needs and work result in publication *"Effect of Shod Walking on Plantar Pressure with Varying Insole"*. His area of research is Design Thinking, Ergonomics for Podiatric Medicine, 3D printing and User Experience Design. He is also the editor of various books: *Emerging Application of 3D Printing during COVID-19, Application of 3D Printing in Biomedical Engineering, Revolutions in Product design for Health care, 3D printing in Podiatric Medicine and Food Printing: 3D printing in Food Sector.* He has also acting as an editorial review board member for *International Journal of Technology and Human Interaction* (IJTHI), *Advances in Science, Technology and Engineering Systems Journal* (ASTESJ) and also a review editor for frontiers in manufacturing section "Additive Processes". He has established a research collaboration with Prof. Karupppasamy Subburaj at Singapore University of Technology and Design, SINGAPORE on Medical device design and biomechanics.

**Sunpreet Singh** is a researcher in NUS Nanoscience and Nanotechnology Initiative (NUSNNI). He has received Ph.D. in Mechanical Engineering from Guru Nanak Dev Engineering College, Ludhiana, India. His area of research is additive manufacturing and application of 3D printing for development of new biomaterials for clinical applications. He has contributed extensively in additive manufacturing literature with publications appearing in *Journal of Manufacturing Processes, Composite Part: B, Rapid Prototyping Journal, Journal of Mechanical Science and Technology, Measurement, International Journal of Advance Manufacturing Technology,* and *Journal of Cleaner Production.* He authored 150 research papers and 27 book chapters. He is working with joint collaboration with Prof. Seeram Ramakrishna, NUS Nanoscience and Nanotechnology Initiative, and Prof. Rupinder Singh, Manufacturing Research Lab, GNDEC, Ludhiana. He is also the editor of three books:

*Current Trends in Bio-manufacturing*, Springer Series in *Advanced Manufacturing*, Springer International Publishing AG, Gewerbestrasse 11, 6330 Cham, Switzerland, December 2018; *3D Printing in Biomedical Engineering*, book series *Materials Horizons: From Nature to Nanomaterials*, Springer International Publishing AG, Gewerbestrasse 11, 6330 Cham, Switzerland, August 2019, and *Biomaterials in Orthopaedics and Bone Regeneration-Design and Synthesis*, book series *Materials Horizons: From Nature to Nanomaterials*, Springer International Publishing AG, Gewerbestrasse 11, 6330 Cham, Switzerland, March 2019. He is also the guest editor of three journals: Guest Editor of special issue of "Functional Materials and Advanced Manufacturing", Facta Universitatis, Series: Mechanical Engineering (Scopus Index), Materials Science Forum (Scopus Index), and Special Issue on "Metrology in Materials and Advanced Manufacturing", Measurement and Control (SCI indexed).

**Chander Prakash** is Professor in the School of Mechanical Engineering, Lovely Professional University, Jalandhar, India. He has received Ph.D. in mechanical engineering from Panjab University, Chandigarh, India. His areas of research are biomaterials, rapid prototyping and 3D printing, advanced manufacturing, modelling, simulation, and optimization. He has more than 11 years of teaching experience and six years of research experience. He has contributed extensively to the world in the titanium and magnesium-based implant literature with publications appearing in *Surface and Coating Technology, Materials and Manufacturing Processes, Journal of Materials Engineering and Performance, Journal of Mechanical Science and Technology, Nanoscience and Nanotechnology Letters, Proceedings of the Institution of Mechanical Engineers, Part B: Journal of Engineering Manufacture*. He has authored 60 research papers and 10 book chapters. He is also the editor of three books: *Current Trends in Bio-manufacturing*; *3D Printing in Biomedical Engineering*; and *Biomaterials in Orthopaedics and Bone Regeneration—Design and Synthesis*. He is also the guest editor of three journals: special issue of "Functional Materials and Advanced Manufacturing", Facta Universitatis, Series: Mechanical Engineering (Scopus Indexed), Materials Science Forum (Scopus Indexed), and special issue on "Metrology in Materials and Advanced Manufacturing", Measurement and Control (SCI indexed).

**Karupppasamy Subburaj** is an Assistant Professor in the Pillar of Engineering Product Development (EPD) at Singapore University of Technology and Design (SUTD). He leads an interdisciplinary research team to design and develop medical devices, assistive technologies, image-based quantitative biomarkers, and computing tools (machine learning, artificial intelligence) for diagnosing, monitoring, treating, and potentially preventing musculoskeletal disorders (osteoarthritis/osteoporosis) and disabilities by understanding bio-mechanical implications of those diseases and disabilities. He collaborates with physicians and clinical researchers from Tan Tock Seng Hospital (TTSH), Technical University of Munich (TUM), Singapore General Hospital (SGH), and Changi General Hospital (CGH) to combine research and technical expertise to address real-life clinical problems affecting Asia-Pacific and the World. Before joining SUTD, he did his postdoctoral work in the Musculoskeletal Quantitative Imaging Research (MQIR) laboratory at the University of California San Francisco (UCSF). At UCSF, he worked with a spectrum of clinicians (from radiology, orthopaedic surgeons, sports medicine, and physiotherapy/rehabilitation science) on characterizing magnetic resonance image (MRI) based bio-markers to understand the physiological and biochemical response of knee joint cartilage to physical exercise and acute loading. He has also developed and validated 3D modelling and quantification methods to study joint (hip/knee) loading patterns and contact kinematics in young healthy adults and patients with osteoarthritis. He received his PhD from the Indian Institute of Technology Bombay (IIT Bombay), India, in 2009. During his PhD, he collaborated with Tata Memorial Hospital (TMH), Mumbai, on developing a Surgery Planning System for Tumour Knee Reconstruction. He also worked with orthodontists from local hospitals in Mumbai, India, on designing and developing prostheses and surgical instruments/ guides for reconstructing maxillofacial defects. After his PhD, he worked as a research specialist (surgery planning) in the Biomedical Engineering Technology incubation Centre (BETiC) at IIT Bombay, before moving to UCSF for his post-doctoral studies.

**Seeram Ramakrishna** is a co-director, NUS Nanoscience and Nanotechnology Initiative (NUSNNI). He has received his Ph.D. from the University of Cambridge and is a global leader in electrospinning and nanostructured materials. His research has resulted in approximately 1000 peer-reviewed articles with over 115, 888 citations and an h-index of 160. He has published several books and book chapters. He has been recognized as a Highly Cited Researcher in Materials Science.

# Chapter 1
# Sustainablity for 3D Printing

**Henrique Almeida, Eujin Pei, and Liliana Vitorino**

**Abstract** Since the industrial revolution in the eighteenth century, mankind has focused on industrial and economic dominance in the production of products. Due to evolving technologies, processes and materials, economic dominance has become a key factor for product development. But in the last decade, environmental issues have gained importance and it is a critical issue for current and future industries. In this context, additive manufacturing and digital technologies have allowed us to also create new environmental awareness amongst the industrial and scientific community. From the commercial perspective, economic, marketing and social impact are key issues to be addressed. From the industrial perspective, the design, material and processing parameters are critical aspects. All of these issues will influence the uptake and adoption of Additive Manufacturing while increasing environmental awareness. This chapter will provide a global overview on how Additive Manufacturing (AM) has a huge influence on the environment, while increasing both industrial and commercial benefits to society.

**Keywords** Sustainability · Life cycle assessment of additive manufacturing · 3D printing · Rapid prototyping

H. Almeida (✉) · L. Vitorino
School of Technology and Management, Polytechnic Institute of Leiria, Leiria, Portugal
e-mail: henrique.almeida@ipleiria.pt

H. Almeida
Computer Science and Communication Research Centre, Polytechnic Institute of Leiria, Leiria, Portugal

E. Pei
Brunel Design School, Brunel University London, Uxbridge UB8 3PH, UK

© The Author(s), under exclusive license to Springer Nature Switzerland AG 2022
K. Sandhu et al. (eds.), *Sustainability for 3D Printing*, Springer Tracts
in Additive Manufacturing, https://doi.org/10.1007/978-3-030-75235-4_1

## 1.1  Introduction

The exponential growth of human economic expansion has had a devastating effect on the world's natural resources and consequently on the environment [1]. Kates et al. claims the importance of understanding the fundamental character of inter-actions between nature and society as a new paradigm of sustainability science [2]. Everyone has a responsibility to contribute to a better world. If sustainability becomes a reality, consumer behaviour and perspectives will have to change [3]. Organisations should change to "green" management and develop initiatives in accordance to the circular economy principles and keep in mind that the organi-sation goals are now dual: social and financial [4]. On the other hand, consumers should pay more attention to their attitudes and concerns with environmental issues. Some activities of green consumption behaviours are already in practice and some examples of their behaviour are [5].

- Consumers to avoid those products which have impact on environment;
- Aerosols containing products avoided;
- Prefer recycled based products, e.g. papers;
- Focus on organic foods;
- Foods: Prefer local one;
- Local stores prefer for purchase.

From a marketing point of view, we understand that industries need to recognise this new trend in consumer behaviour to align their marketing strategies with their consumer personal values [6] and eventually change its product program (formula, package, labels). However, some studies showed that consumers that report positive attitudes towards services and eco-friendly products follow through with their wallets revealing a gap between intention and action [7, 8]. *Thus, the role of designer and/or product developer is to keep three pillars (environmental, social and economically) of sustainability in mind while design and development of any product* [9].

Sustainability may be achieved through many different ways, including gov-ernments, new technologies, industries and markets [10]. Additive manufacturing (AM) or 3D printing as it is commonly called, has been recognised as a sustainable and efficient technology. These technologies allow manufacturers to use only the necessary amount of materials, an advantage that can add economic value by reducing both material and production costs [11]. 3D printing also improves operating efficiency by reducing design, production, inventory, store management costs, distribution and transportation, leading to sustainable industrial practices [12, 13]. Additive manufacturing also allows to eliminate supply chain operations associated with productions and requires extra new tools, enabling the repair and remanufacturing of obsolete or damaged tools and dies, eliminate scrap, eliminate the need for tooling and eliminate the use of environmentally hazardous processes

[14–18]. Fruggiero et al. [19] presented a paper on the comparison between subtractive and additive manufacturing of metal parts and they demonstrate within several domains that AM is much more sustainable in comparison to subtractive manufacturing.

Additive manufacturing systems are typically small in size, therefore they can be easily located nearby any existing market, thus reducing transport logistics of products around the world [20]. On the other hand, the raw materials for 3D printing systems are common, thus leading to a net reduction in transportation costs [21]. Several previous studies [15, 16, 22] have defined the carbon footprint reduction for 3D printing technologies.

The studies describes, the main environmental and sustainable benefits from adopting 3D printing technologies are presented:

- In supply chain required efficient and less amount of materials, however, which are available through natural resources.
- Try to avoid energy-consuming and wasteful type manufacturing processes, e.g. Investment casting and cutting fluids machining, e.g. computer numerical control.
- Ability to design higher efficient products with enhanced operational performance that are more effective than conventionally manufactured components by incorporating conformal channels for cooling and heating and gas flow paths and also allowing to produce more complex components that reduce the number of conventional components and their assembly, etc.
- Ability to eliminate fixed asset tooling, allowing for the manufacture to occur at any geographic location close to their customers, reducing transportation costs within the supply chain.
- Parts of lighter weight when used in the transportation industry (automotive, aeronautics, etc.) more focus on fuel efficiency and eliminates the carbon emissions, e.g. ($CO_2$, $CH_4$).

Majeed et al. [23] developed a framework which is based on big data analytics that may be used as a guideline to select related product manufacturing cycle stages that influences the sustainable production of a specified AM system. The results from their work indicate that controlled energy consumption and product's quality are helpful for smart sustainable manufacturing, lower emissions and cleaner productions.

The above benefits allow AM to have the following environmental impacts: lower resource consumption of both energy and material, less waste management and improved pollution control [24]. For a global understanding of the sustainability of additive manufacturing, the next figure will provide its impacts not only on the environment but also on both the economy and society (Fig. 1.1).

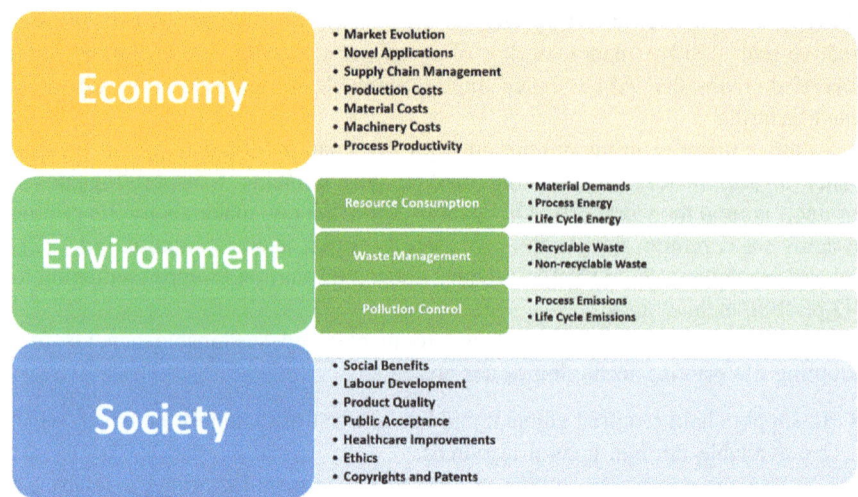

**Fig. 1.1** Sustainability of additive manufacturing (adapted from [24])

## 1.2  Sustainability

### 1.2.1  Design in 3D Printing (Df3DP)/Design for DfAM

The concept of Design for 3D printing (Df3DP) proposes a structured, design-centric approach that combines the use of AM technologies to minimise the manufacturing steps during new product development and its associated processes, yet achieving maximum functionality with a lower unit cost [25]. Df3DP capitalises on the benefits of AM such as realising freedom of geometry, offering customisation, enabling integrated assemblies and aligning towards automated production. Wiberg et al. [26] provided a state-of-art review of the Df3DP research domain in terms of design guidelines, methodologies and available software. They state that Design for 3D printing could be broadly categorised from the perspective of system, process and part design, where system design covers what should be manufactured using AM technologies and the component's boundaries. Process design involves the steps of how the preparations and steps of the component should be performed with clear objectives. Lastly, part design investigates how a single part should be designed and it is recognised as being the most important category.

Vaneker et al. [27] summarise the tools and methods that could be used within Df3DP for designing lightweight parts including the use of Generative Design, Topology Optimization, Lattice Structure Filling, Functional Material Complexity, Internal Geometries, Printed Permeability and Assembly and Part Integration. By considering these aspects for DfAM and subsequent methods, end-users would be able to achieve better efficiency with sustainability in mind. This is in line with Niaki et al. [28] who cited that the use of AM offers several benefits from a sustainability perspective by using less resources [29] and with less operational requirements [30]. Another example is cited by Yang et al. [31] about the use of part consolidation produced via AM with other strategies including high void-to-solid ratio, weight reduction and selecting the most environmentally friendly processes.

## 1.2.2  Designing for 3D Printing

As discussed in the previous section, part design investigates how a single part should be designed within the Df3DP process. Part design should consider design rules for AM as well as process-specific design guidelines for AM. According to Mani et al. [32], design rules consider aspects of design potentials, design restrictions and process capabilities that provide insight into manufacturability. This is extended from the work of Adam and Zimmer [33] who proposed Direct Manufacturing Design Rules (DMDR) where key aspects of such design rules encompass design for function (functional integration, design potentials, etc.), design for tolerance (e.g. physical restrictions) and capabilities (speed, accuracy, repeatability, material, etc.). Mani et al. [32] also proposed Design Principles that are logical aspects derived from guidelines and fundamentals; whereas fundamentals are purely logical primitives that capture process and control parameters. They state that guidelines aim to provide an understanding of AM categories, processes, operating procedures and best practices.

Other more specific guides include those for part orientation [34]; part consolidation [35]; general metal AM [36]; design rules for Selective Laser Melting [37] and design rules for Wire Arc AM [38]. Design guides have also been published by independent machine manufacturers such as EOS GmbH [39], Stratasys Direct Inc. [40] and other service bureaus [41]. Standardisation efforts are also being pursued between ASTM International sub-committee F42.04 and the International Organisation for Standardisation ISO/TC261 to develop joint projects. Currently, standards have been published for ISO/ASTM 52910:2018, ISO/ASTM 52911-1:2019, ISO/ASTM 52911-2:2019 and ISO/ASTM WD 52911-3.

## 1.2.3  Summary

Design for DfAM provides a structured, design-centric approach towards effective use of AM. In addition, Part Level DfAM Methods and other Principle-to-Rule DfAM Methodologies, as well as academic research, publications by independent machine manufacturers, service bureaus and standardisation organisations exist. By being well versed in these guides that offer industry best practices, users can capitalise on the use of AM to their fullest advantage and with sustainability in mind (Table 1.1).

**Table 1.1** Categories and methods for DfAM

| Categories of DfAM [26] | Part level DfAM methods [27] | Principle-to-rule DfAM methodology [32] |
| --- | --- | --- |
| System level | Topology optimization | Design guidelines (States AM categories, processes, operation and best practices) |
| Process level | Generative design | Design rules (States potentials and constraints based on principles) |
| Part level | Lattice structure filling | Design principles (States logical principles based on fundamentals) |
| | Internal geometries | Design fundamentals (States process and control parameters) |
| | Printed permeability | |
| | Functional material complexity | |
| | Assembly and part integration | |

## 1.3   Sustainability and 3D Printing Processes

Over the past few years, many researchers have gained awareness of sustainability in 3D printing processes. Some 3D printing processes have had higher focus of research than others due to their specific characteristics. But, in spite of their differences, there is a global similarity between all 3D printing systems. Determining the optimal orientation of the part for production is both a time consuming and difficult task since one has to trade-off numerous contradicting objectives such as production time and the part's surface finish [20, 42–45]. An inappropriate choice results in physical model with a significant "staircase effect" resulting in parts with poor surface quality [46] or excess production time consuming unnecessary material and/or energy.

By consuming only the necessary amount of raw material for building the desired part, AM improves the material management for part production reducing both the energy consumption (not always) and the amount of waste material and when compared to other conventional technologies, also eliminates scrap, the need for tooling and the use of environmentally hazardous process enablers [14, 15, 18]. In some cases, the energy consumption might be higher due to the moderately low productivity and the use of elevated power and temperature processing. Mognol et al. [47] presented a study using the design of experiments methodology for energy consumption reduction considering the following parameters: Layer thickness, part orientation and part position in the build chamber. No specific energy model or optimization rules for time reduction were developed despite the study undergoing on three AM systems. Energy consumption in 3D printing processes are therefore directly related with its classification. According to ASTM International, the existing AM technologies are classified as follows [20, 48]:

1. Material extrusion—process that creates layers by mechanically extruding molten thermoplastic material onto a platform.
2. Powder bed fusion—these techniques use an energy beam, either a laser or electron beam, to selectively melt a layer of powder material.
3. Vat photopolymerization—an ultraviolet laser is used to selectively polymerize a UV curable photosensitive resin creating a layer of solidified material which are subsequently cured until the part is complete.
4. Material jetting—these techniques directly deposit wax or photopolymer droplets onto a substrate via drop-on-demand ink jetting.
5. Binder jetting—this process deposits a stream of particles of a binder material over the surface of a powder bed, joining particles together creating the object.
6. Sheet lamination—layers of adhesive-coated paper or plastic are successively glued together and then cut to shape with a knife or laser cutter.
7. Directed energy deposition—metallic powder or wire is fed directly into the focal point of an energy beam creating a molten pool of material to build the part.

Considering the above classification, one may divide the 7 categories into 3 categories according to their energy consumption, namely:

- Low energy: Material extrusion, Binder jetting, Material jetting, Sheet lamination
- Medium energy: Vat photopolymerization, Material jetting (photopolymer requiring UV light), Sheet lamination (laser cutter)
- High energy: Powder bed fusion, Directed energy deposition

Another aspect to be considered during production is the fabrication of support structures. An inappropriate orientation results in excessive building of support structures around the physical part or creation of supports within specific areas of the desired part, which are difficult or almost impossible to remove, increasing significantly the energy and effort of removal of the support structures [20, 49].

Almeida and Correia presented a work on support structures for material extrusion systems where they studied: (1) the relationship between the volumes of support material used in material extrusion systems and the time necessary to dissolve the support material (2) the environmental impact of different support material schemes for embracing the parts during production [50]. In the first case, the relationship is directly proportional, but in the second case, significant differences were encountered. A normal material extrusion system has a software that allows to define the type of support structure scheme in order to embrace the desired part during production. Most commonly, the software system has the following support structure schemes: Smart, Sparse, basic and Surround. They demonstrated that each one of the support structure schemes influences more than just increased production time which is one of the main criteria. Other issues are the amount of support material used during production, the energy consumption during production, the energy consumption during the support structure removal and the usage of the amount of dissolution liquid for the support structure removal. The comparison was not only performed in hours [h] and kilowatt-hours [kwh], but also in energy consumption impacts [mPT] for ecodesign purposes. The authors also state that these variables are highly influenced by the design of the parts been produced, meaning that in some cases, the differences may have slight significance but in others, the differences are significant.

A systematic review of supports structures for AM was performed by Jiang et al. [51] where they discuss several methods of support structure design for minimising the use of support structures without compromising the part's fabrication. The authors also present suggestions of design modifications of the desired part that will reduce the need for support structures. By reducing the amount of material, time and energy involved, the process becomes more sustainable while producing the desired part or component.

Materials for AM has also been subject to sustainable research [52]. Vidakis et al. [53] studied the mechanical performance of recycling acrylonitrile–butadiene–styrene (ABS) polymer on extrusion-based systems. Standard tensile, compression, flexion, impact and micro-hardness tests were performed on each ABS filament on each recycle repetition for six recycling cycles, evaluating the effect of the thermomechanical treatment on the ABS material during each recycling process. The authors demonstrated that the mechanical performance of the recycled ABS material improved for each recycle repetition for a certain number of repetitions. An optimal mechanical behaviour was determined between the 3rd and 5th recycling repetition, indicating a positive impact of the ABS material recycling for extrusion-based systems, contributing towards improved sustainability.

Daraban et al. [54] presented a review regarding the use of metals for additive manufacturing. Results from their literature review indicate that metal powder recycling and use/reuse technologies could be developed to economise metal powder and the use of metallic AM in existing component redesign and repairs also increase sustainability. The authors also state that the sustainable performance of metallic AM technologies depends on the quality of the metal powder and its lifecycle. Therefore, metal powder recovery and recycling optimisation is both a major research topic and an industrial necessity.

Regarding the AM processes more specifically, several authors have performed different studies to analyse its sustainable impact. For instance, for SLS systems, Sreenivasan and Bourell [15, 55] performed a quantitative energy calculation regarding several systems, namely the laser system, heater, roller drives, piston control and other miscellaneous systems within the SLS system. They determined that the heating system was the principal energy consumer, followed by the drives and controls and finally the laser system [15]. During their study, the SLS system worked with average power and no relationship with process parameters was discovered.

Fredriksson [56] investigated the material and manufacturing life-cycle stages of INCONEL 718. Energy measurements from an ARCAM A2X Electron Beam Melting system was determined and compared to the embodied energy and indirect $CO_2$ emissions of the feedstock as well as conventional subtractive manufacturing. Fredriksson [56] demonstrated that the production of the metal powder and the AM process itself contributes considerably to the total energy usage and emissions.

The author used the Ashby's 5-step method for the assessment of sustainable development to briefly discuss the social and economic impacts of additive technologies.

Fargione and Giudice [57] proposed a sustainable oriented DfAM approach, to analyse the dependence of the energy impact on the geometric characteristics and the material ($Ti_6Al_4V$) of an Electron Beam Melting system. The quantification of the environmental impact of the built parts focused on the determination of energy consumption of the additive process and correlating it to the main process parameters and also to some features that characterise the shape of the part. The authors developed a model for quantifying energy consumption for both the system and material allowing for a direct control regarding energy sustainability, focused on design variable choices and process parameter settings.

Majeed et al. [23] developed a framework based on big data analytics that can be used as a guideline to select the related product manufacturing cycle stages that influence the sustainable production of a specified AM system. The results from their work indicate that the quality of the part and energy consumption are adequately controlled which is helpful for smart sustainable manufacturing, lower emissions and cleaner productions.

Concerning binder jetting processes, Meteyer et al. [58] presented a material and energy consumption model as a function of process parameters and part geometry and later continued their work focusing on the building stage, where both the layer thickness and part orientation was studied [59].

Freitas et al. [60] studied the production of several parts as a function of part internal filling and building orientation, regarding energy consumption, production time and end-of-life scenarios for extrusion-based systems. The authors used eco-indicators and enabled them to compare the environmental impact of the product's material and the energy consumption of extrusion-based systems. They also state that in order to reduce the environmental impact of energy consumption, it is vital to shorten the time the machine is on without production, as well as reducing the amount of productions by combining more parts during production, diluting the pre-heating and cooling of the machine between productions. No specific model was obtained, but by combining life-cycle assessment with production parameters, a better awareness was provided towards the users of extrusion-based systems. Moreover, authors used 3D printed tool to cut soft material and try to minimise the high cost tool for machining [61, 62].

Regarding stereolithography-based systems, Yang et al. [63] presented a mathematical model for the energy consumption. In order to validate their model, experiments to measure the real energy consumption of the SLA system was conducted. A design of experiments method was implemented to study the impacts of the different processing parameters and their potential interactions on the energy consumption. A response optimization method was used to recognise the optimum combination of parameters in order to minimise the total energy consumption. The results demonstrate that the global energy consumption of the SLA system can be significantly reduced with optimum parameter settings without visible quality decay of the produced parts.

## 1.4    Conclusions

Since the industrial revolution in the eighteenth century, mankind has focused on industrial and economic dominance in the production of products. Due to evolving technologies, processes and materials, economic dominance has become a key factor for product development. But in the last decade, environmental issues have gained importance and is a critical issue for current and future industries.

Sustainability allows to create and maintain the conditions for humans and nature to coexist in a productive harmony, fulfilling the social, economic and other requirements of present and future generations. Environmental and social worries about the human's impact on the environment have pushed the development of sustainable issues. Sustainable industrial practices contribute to the development of more sustainable materials, products and processes. It is critical to apply eco-design principles and develop greener production processes and products, reducing the ecological impacts associated with both production and product consumption.

In this context, AM and digital technologies have allowed us to also create new environmental awareness amongst the industrial and scientific community. From a business perspective, economic, marketing and social impact are key issues to be addressed. From a production perspective, the design, material and processing parameters are critical aspects. All of these issues will influence the uptake and adoption of AM while increasing positive environmental impacts and industrial and commercial benefits to society.

## References

1. Chams, N., García-blandón, J.: On the importance of sustainable human resource management for the adoption of sustainable development goals. Resour. Conserv. Recycling **141**(Sept 2018), 109–122 (2019). https://doi.org/10.1016/j.resconrec.2018.10.006
2. Kates, R.W., Clark, W.C., Corell, R., Hall, J.M., Jaeger, C.C., Lowe, I., Mccarthy, J.J., Schellnhuber, H.J., Bolin, B., Dickson, N.M., Faucheux, S., Gallopin, G.C., Grubler, A., Huntley, B., Jager, J., Jodha, N.S., Kasperson, R.E., Mabogunje, A., Matson, P., ..., Svedin, U.: Sustainability Science. Science **292**(5517), 641–642 (2001)
3. Vitorino, L., Lisboa, A., Antunes, R.: Digital era: how marketing communication develops business innovation—case studies. In: Digital Marketing Strategies and Models for Competitive Business, ch. 1 (2020). https://doi.org/10.4018/978-1-7998-2963-8.ch001
4. Harris, C., Tregidga, H.: Hr managers and environmental sustainability: strategic leaders or passive observers? Int. J. Hum. Resour. Manage. **23**(2), 236–254 (2012)
5. Gilg, A., Barr, S., Ford, N.: Green consumption or sustainable lifestyles? Identifying the sustainable consumer. Futures **37**(6), 481–504 (2005). https://doi.org/10.1016/j.futures.2004.10.016
6. Rosmarin, R.: Sustainability sells: why consumers and clothing brands alike are turning to sustainability as a guiding light. Business Insider (2020)
7. White, K., Hardisty, D.J., Habib, R.: The Elusive Green Consumer. Harvard Business Review (2019). https://hbr.org/2019/07/the-elusive-green-consumer

8. Orzan, G., Cruceru, A.F., Balaceanu, C.T., Chivu, R.G.: Consumers' behavior concerning sustainable packaging: an exploratory study on Romanian consumers. Sustainability (Switzerland) **10**(6), 1787 (2018). https://doi.org/10.3390/su10061787

9. Diegel, O., Kristav, P., Motte, D., Kianian, B.: Design, additive manufacturing and its effect on sustainable. In: Muthu, S.S., Savalani, M. (eds.) Handbook of Sustainability in Additive Manufacturing, vol. 1, pp. 73–100. Springer (2016)

10. Antonides, G.: Sustainable consumer behaviour: a collection of empirical studies. Sustainability **9**(1686), 1–5 (2017)

11. Morand, P.: Op-Ed: what 3D printing means for fashion. Retrieved from https://www. businessoffashion.com/articles/opinion/3d-printing-technology-disrupt-fashion-and-luxury-pascal-morand (2016). Accessed 23 Oct 2020

12. Chabaud, C.: 3D printing and the future shape of retail industry. Retrieved from https://www. sculpteo.com/blog/2015/12/23/3d-printing-and-retail-industry/ (2015). Accessed 23 Oct 2020

13. Kilbert, A.: 3D printers could revolutionize fashion industry. Retrieved from http://www. duqsm.com/3d-printers-could-revolutionize-fashion-industry/ (2016). Accessed 23 Oct 2020

14. Gebler, M., Uiterkamp, A.J.M.S., Visser, C.: A global sustainability perspective on 3D printing technologies. Energy Policy **74**, 158–167 (2014)

15. Sreenivasan, R., Goel, A., Bourell, D.L.: Sustainability issues in laser-based additive manufacturing, Phys. Procedia **5**(Part A), 81–90 (2010)

16. Reeves, P.: Additive Manufacturing—A supply chain wide response to economic uncertainty and environmental sustainability. In: International Conference on Industrial Tools and Material Processing Technologies, Ljubljana, Slovenia (2009)

17. Morrow, W.R., Qi, H., Kim, I., Mazumder, J., Skerlos, S.J.: Environmental aspects of laser-based and conventional tool and die manufacturing. J. Clean. Prod. **15**, 932–943 (2007)

18. Hague, R.: Unlocking the design potential of rapid manufacturing. In: Hopkinson, N., et al. (eds.) Rapid Manufacturing: An Industrial Revolution for the Digital Age. Wiley (2005)

19. Fruggiero, F., Lambiase, A., Bonito, R., Fera, M.: The load of sustainability for additive manufacturing processes. Procedia Manuf. **41**, 375–382 (2019)

20. Gibson, I., Rosen, D., Stucker, B.: Additive manufacturing technologies—3D printing, rapid prototyping and direct digital manufacturing, 2nd edn. Springer, New York (2015)

21. Gibson, I.: Is additive manufacturing a sustainable technology?. In: Bártolo, H., et al. (eds.) Proceedings of SIM2011 Sustainable Intelligent Manufacturing. IST Press, pp. 583–589 (2011)

22. Bourell, D.L., Leu, M.C., Rosen, D.W.: Roadmap for additive manufacturing: identifying the future of freeform processing. The University of Texas at Austin (2009)

23. Majeed, A., Zhang, Y., Ren, S., Lv, J., Peng, T., Waqar, S., Yin, E.: A big data-driven framework for sustainable and smart additive manufacturing. Robot. Comput. Integr. Manuf. **67**, 102026 (2021)

24. Peng, T., Kellens, K., Tang, R., Chen, C., Chen, G.: Sustainability of additive manufacturing: an overview on its energy demand and environmental impact. Addit. Manuf. **21**, 694–704 (2018)

25. Joo, S.H., et al.: What is design for additive manufacturing (DfAM)?. In: Additive Manufacturing Applications for Metals and Composites. IGI Global, pp. 164–186 (2020)

26. Wiberg, A., et al.: Design for additive manufacturing—a review of available design methods and software. R. Prot. J. **25**(6), 1080–1094 (2019)

27. Vaneker, T., et al.: Design for additive manufacturing: framework and methodology. CIRP Ann. Manuf. Technol. **69**(2), 579 (2020)

28. Niaki, M.K., et al.: Why manufacturers adopt additive manufacturing technologies: the role of sustainability. J. Clean. Prod. **222**, 381–392 (2019)

29. Ullah, A.S., et al.: Sustainability analysis of rapid prototyping: material/resource and process perspectives. Int. J. Sustain. Manuf. **3**(1), 20–36 (2013)

30. Weller, C., et al.: Economic implications of 3D printing: market structure models in light of additive manufacturing revisited. Int. J. Prod. Econ. **164**, 43–56 (2015)

31. Yang, S., et al.: Understanding the sustainability potential of part consolidation design supported by additive manufacturing. J. Clean. Prod. **232**, 722–738 (2019)
32. Mani, M., et al.: Design rules for additive manufacturing: a categorization. In: International Design Engineering Technical Conferences and Computers and Information in Engineering Conference, vol. 58110, ASME (2017)
33. Adam, G.A., Zimmer, D.: Design for additive manufacturing—element transitions and aggregated structures. CIRP J. Mfg. Sci. Tech. **7**(1), 20–28 (2014)
34. Leutenecker-Twelsiek, B., et al.: Considering part orientation in design for additive manufacturing. Procedia CIRP **50**, 408–413 (2016)
35. Liu, J.: Guidelines for AM part consolidation. Virt. Phys. Prot. **11**(2), 133–141 (2016)
36. Samperi, M.T.: Development of design guidelines for metal additive manufacturing and process selection. Master Thesis, Pennsylvania State University (2014)
37. Thomas, D.: The development of design rules for selective laser melting. Doctoral dissertation, University of Wales (2009)
38. Lockett, H., et al.: Design for wire + arc additive manufacture: design rules and build orientation selection. J. Eng. Des. **28**(7–9), 568–598 (2017)
39. EOS GmbH.: Design Rules for DMLS. http://www.3dimpuls.com/sites/default/files/download/dmls_design-rules_en.pdf (2015). Accessed 12/09/2020
40. Stratasys Direct Inc.: Laser Sintering Design Guidelines. https://stratasysdirect.com/resources/laser-sintering (2020). Accessed 12/09/2020
41. Crucible Design Ltd.: Design Guidelines for Direct Metal Laser Sintering (DMLS), Abingdon, Oxfordshire, UK. https://www.crucibledesign.co.uk/guides/bs7000-part-2-a-management-guide.php. (2020). Accessed 12/09/2020
42. Rosen, D.: What are the principles for design for additive manufacturing? In: Kai, C.C., et al. (eds.) Proceedings of the 1st International Conference on Progress in Additive Manufacturing (Pro-AM2014), Research Publishing Services, pp. 85–90 (2014)
43. Pham, D.T., Demov, S.S.: Rapid Manufacturing: The Technologies and Applications of Rapid Prototyping and Rapid Tooling. Springer, London (2001)
44. Chua, C.K., Fai, L.K.: Rapid Prototyping: Principles and Applications in Manufacturing. World Scientific (2000)
45. Alexander, P., Allen, S., Dutta, D.: Part orientation and build cost determination in layered manufacturing. Comput. Aided Des. **30**(5), 343–356 (1998)
46. Thrimurthulu, K., Pandey, P.M., Reddy, N.V.: Optimum part deposition orientation in fused deposition modeling. Int. J. Mach. Tools Manuf. **44**, 585–594 (2004)
47. Mognol, P., Lepicart, D., Perry, N.: Rapid prototyping: energy and environment in the spotlight. Rapid Prototyp. J. **12**(1), 26–34 (2006)
48. Gao, W., Zhang, Y., Ramanujan, D., Ramani, K., Chen, Y., Williams, C.B., Wang, C.C.L., Shin, Y.C., Zhang, S., Zavattieri, P.D.: The status, challenges, and future of additive manufacturing in engineering. Comput. Aided Des. **69**, 65–89 (2015)
49. Chua, C.K., Leong, K.F.: 3D Printing And Additive Manufacturing—Principles and Applications, 4th edn. World Scientific Publishing (2014)
50. Almeida, H.A., Correia, M.S.: Sustainable impact evaluation of support structures in the production of extrusion-based parts. In: Muthu, S., Savalani, M. (eds.) Handbook of Sustainability in Additive Manufacturing. Environmental Footprints and Eco-design of Products and Processes. Springer, Singapore (2016). https://doi.org/10.1007/978-981-10-0549-7_2
51. Jiang, J., Xu, X., Stringer, J.: Support structures for additive manufacturing: a review. J. Manuf. Mater. Process. **2**(4), 64 (2018). https://doi.org/10.3390/jmmp2040064
52. Colorado, H., Velásquez, E., Monteiro, S.: Sustainability of additive manufacturing: the circular economy of materials and environmental perspectives. J. Market. Res. **9**(4), 8221–8234 (2020)
53. Vidakis, N., Petousis, M., Maniadi, A., Koudoumas, E., Vairis, A., Kechagias, J.: Sustainable additive manufacturing: mechanical response of Acrylonitrile-Butadiene-Styrene over multiple recycling processes. Sustainability **12**, 3568 (2020)

54. Daraban, A., Negrea, C., Artimon, F., Angelescu, D., Popan, G., Gheorghe, S., Gheorghe, M.: A deep look at metal additive manufacturing recycling and use tools for sustainability performance. Sustainability **11**(19), 5494 (2019). https://doi.org/10.3390/su11195494
55. Sreenivasan, R., Bourell, D.L.: Sustainability study in selective laser sintering—an energy perspective. In: Proceedings of the 20th Annual International Solid Freeform Fabrication Symposium, SFF 2009 (2009)
56. Fredriksson, C.: Sustainability of metal powder additive manufacturing. Procedia Manuf. **33**, 139–144 (2019)
57. Fargione, G., Giudice, F.: An approach to design for environmental sustainability of additive manufactured metal components. Procedia Struct. Integrity **24**, 758–763 (2019)
58. Meteyer, S., Xu, X., Perry, N., Zhao, Y.F.: Energy and material flow analysis of binder-jetting additive manufacturing processes. Procedia CIRP **15**, 19–25 (2014)
59. Xu, X., Meteyer, S., Perry, N., Zhao, Y.F.: Energy consumption model of binder-jetting additive manufacturing processes. Int. J. Prod. Res. **53**(23), 7005–7015 (2015)
60. Freitas, D., Almeida, H.A., Bártolo, H., et al.: Sustainability in extrusion-based additive manufacturing technologies. Prog. Addit. Manuf. **1**, 65–78 (2016). https://doi.org/10.1007/s40964-016-0007-6
61. Sandhu, K., Singh, G., Singh, S., Kumar, R., Prakash, C., Ramakrishna, S., Królczyk, G., Pruncu, C.I.: Surface characteristics of machined polystyrene with 3D printed thermoplastic tool. Materials **13**(12), 2729 (2020)
62. Singh, S., Singh, G., Sandhu, K., Prakash, C., Singh, R.: Investigating the optimum parametric setting for MRR of expandable polystyrene machined with 3D printed end mill tool. Mater. Today: Proc. **33**, 1513–1517 (2020)
63. Yang, Y., Li, L., Pan, Y., Sun, Z.: Energy consumption modeling of stereolithography based additive manufacturing toward environmental sustainability. J. Ind. Ecol. **21**, S168–S178 (2017). https://doi.org/10.1111/jiec.12589

# Chapter 2
# Biomaterials Printing for Sustainability

**Guravtar Singh, Raja Sekhar Dondapati, and Lakhwinder Pal Singh**

**Abstract** 3D printing (3DP) is one of the emerging technology in 21th century and popular in academics and industries. Wide range of materials printed by 3DP and it should be noted that the flexibility of 3DP materials comes from the range of 3DP systems and the seven categories specified in the ISO/ASTM standard. Also, it is not surpassed by all new printers or processes for novel materials. 3DP can never be used as a stand-alone process, being an integral part of a multi-process system or an optimised multi-system process to suit the production of innovative materials and new product requirements. 3DP technology has gained popularity in all fields such as automobile, healthcare, aerospace, recreation, textiles, apparel and the fashion market, etc. Researchers, textile technologists, designers of apparel, suppliers and distributors have since the last decade; they have been working on implementing 3DP technology in their respective fields. 3DP has been recognised as an innovative and efficient process but still, a lot of work is going on these days to use sustainable materials in 3DP from an environmental viewpoint. This chapter provides a brief review on biomaterials that can be used in this technology for sustainable development and strengthen the knowledge bank to the young researchers working in this field.

**Keywords** Biomaterials · Additive manufacturing · Sustainability · Life cycle assessment

G. Singh (✉) · R. S. Dondapati
School of Mechanical Engineering, Lovely Professional University, Phagwara 144411, India
e-mail: guravtar.14443@lpu.co.in

L. P. Singh
Department of Industrial Engineering, Dr. B R Ambedkar National Institute of Technology Jalandhar, Jalandhar 144011, India

## 2.1  Introduction

Many new materials, such as nanomaterial's, functional materials, biomaterials, smart materials or even fast-drying concrete, have been explored for 3D printability and the use as feed materials for printing specific application parts in recent years due to increasing demand for both product complexity and multi-functionality [1–3]. However, minimal reviews of the recent development of these novel materials and 3D printing applications are available as such [4–7]. When combining novel materials with 3DP, now is the time to answer a significant issue, i.e. is there a sufficient match between these novel materials and present 3DP to fulfil the latest product requirements? Therefore, the major objective of this analysis is to critically examine the fundamental aspects of biomaterials in terms of basic resources for sustainability [7–11]. In fact, the progress in this field depended primarily on the ability to synthesise nanoparticles. The printing using biomaterials can composites incorporate the matrix into the complex structures and positions them effectively. And reinforcements with biomaterials to achieve a device with structural or functional properties that are more useful and cannot be accomplished by either of the constituent alone. The use of reinforcements of particles, fibres or nanomaterial's in polymeric matrices makes PMC manufacturing, which is distinguished by high mechanical efficiency and outstanding functionality [12–19] Biodegradable reinforcements for the development of high-efficacy PMCs are the most commonly performed due to the concerns for saving the natural, therefore governments and companies are being pushed by various activists and customers to pay more attention to the ecological effects of the commodity they purchase. Therefore 3DP using sustainable materials for industrial application is also important. A life cycle assessment (LCA) is a useful tool for quantifying the effect of the climate over the entire life cycle product is also taken into consideration [6] (Table 2.1).

## 2.2  Need of Biomaterials in 3DP

The rapid consumption of petroleum-based polymeric systems and visible polymeric systems in the current era and the waste linked to industrial technology has also received a lot of criticism, therefore a lot of work is being done in the area of sustainable materials [20–23]. Environmental consciousness has achieved a new level of practise with a high focus on the use of naturally driven activities instead of synthetic materials. Moreover, sustainable materials used for 3DP which are using recyclable materials and it involves the use of life cycle assessments helps in reducing carbon footprints, therefore, play an integral role in healthy societies and communities [24–27]. Natural resources, such as oil, coal, drinking water, clean air, fertile soil, rare metals, minerals, etc. are depleting therefore for human survival and to keep the global economy running, wood and biodiversity are important. Therefore, a new class of reinforcements has been launched [28–31]. The word

**Table 2.1**  List of the 3DP processes

| Process category | Description | Technologies | Materials |
| --- | --- | --- | --- |
| Powder bed fusion | Regions of a powder bed are selectively fused by thermal energy | EBM, SLS, SLM,DMLS, SHS | Metals, polymers |
| Direct energy deposition | Focused thermal energy is used to fuse materials by melting as the material is being deposited | LMD, DALM, DMD, LDD | Metals |
| Material extrusion | Material is selectively extruded through a nozzle or orifice | FDM | Polymer-based materials |
| Vat photopolymerization | Liquid photopolymer in a vat is selectively cured by light-activated or ultraviolet polymerization | SLA, DLP | Photo-polymers |
| Binder Jetting | A liquid bonding agent is selectively deposited to join powder materials, followed by an optional final curing process | BJ, PBIH, PP | Polymers, metals, sand |
| Material jetting | Droplets of build material are selectively deposited | MJM, PolyJet, MultiJet, etc | Polymers, waxes |
| Sheet lamination | Sheets of material are bonded to form an object | LOM, UC | Metals, paper |

"Green Composites" came into being, referring to materials that support business, natural environment and end-customers". Natural organic fillers, wood flour and fibres are the most commonly used due to its simplicity and cost-effectiveness to obtain from sawmill waste and provides outstanding thermo-mechanical properties. Therefore the focus is shifted towards the plant fibre for the use in packaging applications [32–35]. The major reason is due to high abundance of these fibres, low weight, high strength, rigidity and biodegradability therefore these fibres can play an important role in sustainable packaging [36, 37]. Furthermore, with the exception of fibrous reinforcements, it is extremely difficult for producers to monitor the distribution or placement of fibrous reinforcements. Inside the polymeric structures, the fillers. Nonetheless, the manufacture of tools itself consumes noticeable materials. Labour and machining leading to high costs and long lead times. The use of 3DP technology with these materials has demonstrated important in cost and lead time reductions while offering various other benefits, such as enormous benefits independence of design and rapid iteration, almost irrespective of the complexity of pieces [38]. In addition, it also offers the chance to manufacture lightweight tools, simplifying transportation and storage operations inside production facilities directly. 40 Improvement in the production of composite 3D printers has motivated the development of pre-blended materials within order to achieve special characteristics and qualities, fillers such as nanoparticles, carbon nanotubes, fibres and grapheme Capabilities. Many benefits could be based on layer-by-layer production, for example, material waste is obtained through 3DP. Important in decreases [39] the supporting materials, which is used to make it possible for increasingly exotic geometries, After 3DP, it can be recycled into raw material. Many CNC machined components are generated from the block, which is often significantly more than the portion to be manufactured, it is said that the waste ratios of materials are with the end of the material in the final section, it can reach up to 19:11 [40]. While it is possible to recycle the waste materials, it will the second consumption of energy and resources is also created as pollution for the atmosphere. In addition to conventional 3DP releases less carbon dioxide in the production process. Compared to conventional energy use, less energy consumption is attributed to factory output patterns and delivered to the warehouse. Apart from material and energy considerations, in the warehouse the life cycle of the processes of 3DP, transportation and the carbon footprint of production can be minimised over the course of the design level, since the processes of production will be once the product structure is practically verified, Stated. The designer may change the strategy for the design in the stage of design and thus help to minimise the carbon footprint until supplying it to the processing sector [41–43]. Additionally, the environmental effect of scaling back or scaling back can be further minimised by the elimination of dynamic supply chains [44]. Kreiger et al. [45] have suggested that the accumulated demand for energy polymer processing goods can be reduced by 41–64% with current 3D open-source low-cost printers. Besides that, transport and rough processing of casting, forging and rough machining Packing's that are important in conventional packaging Processes for production can be streamlined or shortened in 3DP techniques are also appropriate for 3DP to take place. Less environmental impact is made and application of 3DP and its drawbacks mentioned in Table 2.2.

## 2.3   Bio-based materials in 3DP for Sustainability

We incorporate a multitude of elements to establish a paradigm shift in the way medical devices are produced and distributed by integrating the two distinct elements of 3DP and biomaterials [45]. 3D printers facilitate on-demand manufacturing, rapid product creation and customised design, in which combining these elements with biomaterials enables sustainable and renewable goods to be manufactured locally that, could enhance LMICs' social, economic and human health. Another form of thermoplastic filament, in line with the theme of this novel, which bridges the gap between technical advances and human health [46]. The ability to use a wider variety of thermoplastic filaments has recently been recognised as a result of continuous developments in 3DP technology with the materials which have sustainability, renewability and are known as 'bio-based' materials as thermoplastic filaments such as polylactic acid (PLA) has garnered a lot of interest [47–52] The use of natural polymers that show similar material properties to their petroleum-based counterparts in the manufacture of 3D objects with natural

**Table 2.2**  Application of 3DP and its drawbacks

| Methods | Materials | Applications | Benefits | Drawbacks |
|---|---|---|---|---|
| Fused deposition modelling | Continues filaments of thermoplastic polymers Continuous fibre-reinforced polymers | Rapid prototyping toys, advanced composite parts | Low cost, high speed, Simplicity | Weak mechanical properties limited materials (only thermoplastics), layer-by-layer finish |
| Powder bed fusion (SLS, SLM, 3DP) | Compacted fine powders Metals, alloys and limited polymers (SLS or SLM) ceramic and polymers (3DP) | Biomedical electronics, Aerospace | Fine resolution, high quality | Slow printing, expensive, high porosity in the binder method (3DP) |
| Inkjet printing and contour crafting | A concentrated dispersion of particles in a liquid (ink or paste) ceramic, concrete and soil | Biomedical, large structures buildings | Ability to print large structures, quick printing | Maintaining workability, coarse resolution, lack of adhesion between layers, Layer-by-layer finish |
| Stereolithography | A resin with photo-active monomers hybrid polymer-ceramics | Biomedical prototyping | Fine resolution, high quality | Very limited materials, slow printing |
| Direct energy deposition | Metals and alloys in the form of powder or wire Ceramics and polymers | Aerospace retrofitting, repair cladding biomedical | Reduced manufacturing time and cost | Low accuracy, need for a dense support structure |
| Laminated object manufacturing | Polymer composites ceramics paper metal-filled tapes metal rolls | Paper manufacturing, foundry industries electronics | Reduced tooling and manufacturing time | Inferior surface quality and dimensional accuracy |

bio-based materials are being encouraged for sustainability and biocompatibility complexes. Study in the field of bio-based materials and derived polymers is due to the growing environmental and sustainability issues associated with conventional petroleum-based polymers such as ABS plastics [53]. In the fields of medicine, biomedical engineering and polymer chemistry in particular, an inherent drive has been fostered to encourage the use of bio-based materials that exhibit the same modular material properties as petroleum-based plastics, but are safe and effective to use in many medical settings, particularly in the case of human tissue exposure [54]. Bio-based polymers are derived from entities of organic biomass, such as maize, sugarcane or cellulose. Therefore used these to manufacture bioplastics, which are a type of plastic, bio-based polymers, can be used. Instead of conventional petroleum based, that is derived from biological materials. Not only naturally occurring bio-based polymers include polymeric materials, but also substances that have been polymerized into natural substances materials of high molecular weight by chemical and/or biological methods [55]. This further extends the bio-based polymer constituents to Different synthetic polymers derived from renewable resources and $CO_2$ are included [56]. The biopolymers as polynucleotides, polyamides, polysaccharides, polyoxoesters, Polythioesters, polyanhydrides, polyisoprenoids, polyphenols and their respective properties derivatives respectively [57, 58].

Therefore, these types of polymers are the most ideal printing material for additive manufacturing processes such as FDM due to their low melting temperature. While biopolymers and bio-based plastics really are a plethora that can be extracted from renewable, renewable sources, each material has different chemical and mechanical properties to be regarded. References [59–62], the scope of application is as equally important as the material properties of the bioplastics, particularly when discussing 3DP processes. Materials processability is a core attribute when 3D fabricating devices with bioplastics, as these materials display varying chemical and mechanical profiles including varying extrusion temperatures, micron layer densities, thermal and hydro-degradability, as well as distinct dexterity, tensile, Load-bearing capacities and compression [63]. There are four major biomaterial types that can be used for the development of bioplastics, including: cellulose derivatives, starch-based plastics, polylactic acid (PLA) and polyhydroxy alkanoates (PHAs) [64–66]. Properties and processing capacities, making each one unique in its application and use respectively [67–70].

### 2.3.1 Cellulose

Cellulose is another material that can be used to make bioplastics. The cellulose a high degree of crystallinity is shown in its structure and its related derivatives; it is more durable than starch and cannot be dissolved or plasticized with ordinary starch [71]. For thermoplastic production, cellulose still needs to be changed due to its thermal degradation tendency. The most suitable material for thermoplastic cellulose is cellulose acetate (CA), which is prepared by hydroxyl group acetylation pulp containing acetic anhydride. The mechanical and thermal properties of cellulose acetate, as well as the biodegradability complex, depend on the amount of acetylation it undergoes. Cellulose, even though their monomeric building blocks are very distinct and starch show distinct properties of materials. Starch and cellulose both have a commonality in that, they may not be processed for applications such as 3D, as a thermoplastic material without prior chemical alteration [72, 73]. Poly (3-hydroxybutyrate) or P3HB are rather crystalline and have thermal and mechanical properties that are equivalent to PP and PE. However, this comes at an expense, as P3HB shows slow crystallisation and break elongation is lower compared to PPP, a measure of versatility [74, 75].

## 2.3.2 Polylactic-Acid

The broad variety of bio-based polymers used to manufacture different bioplastic materials with regard to its applications in 3DP and the manufacture of medical devices, one unique sub-group of products has been of great interest. Scientists are currently focusing on developing bio-derived polymer matrix and filler materials for the manufacture of a biodegradable polymer composite [76]. Polylactic acid (PLA) has been listed as one of the most important biodegradable polymers with great potential to replace petroleum-based polymers in the near future. Any such polymer matrix (lactic acid) or polylactide (PLA) has a high potential to replace most traditional polymer matrices. This comes from renewable resources such as wheat, starch and sugar cane [77–79]. PLA is recognised by ring-opening lactide polymerisation, a dimer of lactic acid. Therefore, it should be referred to as poly-lactide rather than polylactic acid. It is an essential biopolymer of semi-crystalline thermoplastic polyester produced from annual renewable resources such as maize, sugar cane, etc. [61, 80–82]. PLA biopolymer's developmental history dates back to 1893 when lactide formulae were published by Bischoff and Walden. In 1932 Carothers and his associates started manufacturing low molecular weight PLA [83]. Japan began industrial PLA fibre production in 1994. In 2002, Cargill Dow LLC, USA, developed the two stereo-isomeric forms of lactic acid PLA [84–86] derived from the starch under the trade name of Nature Works at a capacity of 140,000 tons/year. Recently nature works updated its trade name to industrial PLA as IngeoTM. PLA attracts further attention because of its renewability, biodegradability, bio-compatibility and good mechanical properties [87]. In addition, PLA degradation releases lower-level non-toxic gasses. PLA have, however, limitations on low thermal stability and extreme brittleness (L). The fragility and improvement of PLA's mechanical and thermal properties are therefore needed if we are to be more successful in long-term large-scale applications The PLA polymer belongs to a class of aliphatic polyester, derived from propionic 2-hydroxy acid (i.e. lactic acid). The basic monomer for PLA production is lactic acid which can be produced either through the fermentation of carbohydrates or through the route of chemical synthesis. There are two stereo-isomeric forms of lactic acid that are L (Laevorotatory) —lactic acid and D (Dextrorotatory)—lactic acid both lactic acid types can be made from either petrochemical or agricultural by-products. Lactic acid derived from the route of petrochemical synthesis gives 50:50 ratios of forms of both L-lactide and D-lactide [88]. Unlike the petrochemical path, the lactic acid derived from agricultural by-products mainly exists in a stereo-isomeric form of L-lactic acid. Currently lactic acid used to manufacture PLA comes from agricultural by-products utilising a fermentation process [89]. Lactic acid polymerization can be accomplished utilising two key methods. The first approach involves direct condensation under high temperature and vacuum of the lactic acid. Nevertheless, PLA can only be produced with low molecular weight with this process mainly because of the presence of water as a by-product in this process which is difficult to separate from the polymer [90]. Two steps are involved in the second process by which PLA is generated from lactic acid first, the low molecular weight prepolymer was formed with the polymerisation reaction of the lactic acid condensation. The low molecular weight prepolymer is depolymerized catalytically to form an intermediary lactic acid called dilactide which is the primary feedstock for higher molecular weight PLA processing. Finally, the dilactide is polymerized to create high molecular weight PLA polymer in the solvent-free ring-opening polymerisation (ROP) [91]. Table 2.3 provides measurements of mechanical and physical properties of PLA and other traditional polymer matrices.

Until the last few years, the usage of PLA had primarily concentrated on food, diapers, farming and other sanitary goods.

However, for a variety of applications such as structural, automotive etc., some studies suggested PLA is a potential bio-derived polymer matrix with the addition of fibrous filler. Natural fibres as fillers for the production of PLA polymer composites offer many advantages, such as light weight, lower manufacturing costs, simple handling, good thermal, tensile strength, less shrinkage and acoustic insulation properties [92–94]. To address PLA's drawbacks, various forms of filler materials must be incorporated into the PLA matrix. Natural fibres and natural clay materials are the two highly cost-effective types of fillers that have many benefits such as improved mechanical and thermal structural properties, easy handling, lightweight, reusability, biodegradability and biocompatibility. Thanks to their environmentally friendly properties, natural fibres have identified as one of the best reinforcing fillers in the manufacture of polymer composites over the last few years [95]. Some of the significant characteristics of natural fibres are ready-to-use with low cost, strong mechanical properties, low weight, environmentally friendly,

**Table 2.3** Comparison of physical and mechanical properties of PLA and other commodity polymers

| Properties | PLA | Polystyrene | Polypropylene | Polyethylene terephthalate |
|---|---|---|---|---|
|  |  | PS | PP | PET |
| Density (g/cm$^3$) | 1.24 | 1.04–1.06 | 0.91 | 1.37 |
| Clarity | Transparent | Transparent | Translucent | Transparent |
| Tensile yield strength (MPa) | 48–110 | 34–46 | 21–37 | 47 |
| Tensile modulus (GPa) | 3.5–3.8 | 2.9–3.5 | 1.1–1.5 | 3.1 |
| Tensile elongation (%) | 2.5–100 | 3–4 | 72 | 79 |
| Impact strength (J/m) | 13 |  | 72 | 79 |
| Glass transition temperature (°C) | 60 | 95 | 0 | 75 |
| Melting temperature (°C) | 153 |  | 163 | 250 |
| Heat distortion temperature (°C) | 55–60 | 84–106 | 80–140 | 74–200 |

biocompatible and biodegradable. PLA shows many benefits, the first being that it can be manufactured on commercially available process equipment. Specifically, as PLA printing filament is highly prominent and used from FDM manufacture, it shows a highly compatible nature with thermo-processing, particularly that of 3-dimensional printing. The rate at which it recrystallizes upon cooling from the melt is a significant property of PLA thermoplastics. The rapid crystallisation needed for many short-cycle plastic applications moments, like 3DP [96, 97].

## 2.4   Summary

Design independence, mass-customisation and the right to print the principal advantages of 3D are complex systems with minimal waste. The sustainability assessment for the new technologies is it is important and it will help to provide opportunities to improve for the latest designers of a product. From the literature it can be concluded the number of processing techniques and materials used, there are both positive and negative effects as to the processing methods and materials used. The effect of 3D printing on the world and it is not easy to draw a valid inference. A fair finding is that depending on the environmental effects of 3D printing, each case is different for a better evaluation of the viability of 3D printing, The decision-makers are led quantitatively and better sustainable materials as discussed in this chapter. To achieve the major aim of sustainable manufacturing. Moreover, sustainable packaging with low-cost rapid tool machining also as good scope for researchers [98, 99].

## 2.5   Future Road Maps

Additive manufacturing (AM) has contributed significantly to the recent study and production of difficult biomaterials for prototyping and personalised frameworks. In spite of the advantages of additive production, there are a few disadvantages that would need further research and development.

- Scientific assessment of the environmental effects on 3DP the systems different 3DP, as mentioned before in this article, processes can produce completely different results with regard to their outcomes.
- Ecological effects- Not only is it relative to different ones, the theory of man-ufacturing, but also in the powders themselves. The method of printing is unstable and several variables can influence the final printing process. Ecological effect, including powder, configuration of parameters, Therefore and even the printers themselves, in order to draw a Convincing inference, which can be used quantitatively, such as LCA the environmental implications of 3D printing while modelling.

- Sustainability not only includes environmental issues, it also covers the economic and social aspects of 3DP sustainability evaluation modelling. Life Cycle Cost (LCC) application to analyse the economic advantage of 3DP has given some positive outcomes, but they are mostly carried out separately, not taking the environmental and social aspects into account.

# References

1. Liu, Z., Jiang, Q., Zhang, Y., Li, T., Zhang, H.C.: Sustainability of 3D printing: a critical review and recommendations. In: International Manufacturing Science and Engineering Conference, Vol. 49903, p. V002T05A004. American Society of Mechanical Engineers (2016)
2. Beg, S., Almalki, W.H., Malik, A., Farhan, M., Aatif, M., Alharbi, K.S., Alruwaili, N.K., Alrobaian, M., Tarique, M., Rahman, M.: 3D printing for drug delivery and biomedical applications. Drug Discovery Today **25**(9), 1668–1681 (2020)
3. Zhang, J., Vo, A.Q., Feng, X., Bandari, S., Repka, M.A.: Pharmaceutical additive manufacturing: a novel tool for complex and personalized drug delivery systems. AAPS PharmSciTech **19**(8), 3388–3402 (2018)
4. Alhnan, M.A., Okwuosa, T.C., Sadia, M., Wan, K.W., Ahmed, W., Arafat, B.: Emergence of 3D printed dosage forms: opportunities and challenges. Pharm. Res. **33**(8), 1817–1832 (2016)
5. Levy, G.N., Schindel, R., Kruth, J.P.: Rapid manufacturing and rapid tooling with layer manufacturing (LM) technologies, state of the art and future perspectives. CIRP Ann. **52**(2), 589–609 (2003)
6. Zeidler, H., Klemm, D., Böttger-Hiller, F., Fritsch, S., Le Guen, M.J., Singamneni, S.: 3D printing of biodegradable parts using renewable biobased materials. Procedia Manuf. **21**, 117–124 (2018)
7. Pakkanen, J., Manfredi, D., Minetola, P., Iuliano, L.: About the use of recycled or biodegradable filaments for sustainability of 3D printing. In: International Conference on Sustainable Design and Manufacturing, pp. 776–785. Springer, Cham (2017)
8. Kohtala, C.: Addressing sustainability in research on distributed production: an integrated literature review. J. Clean. Prod. **106**, 654–668 (2015)
9. McAlister, C., Wood, J.: The potential of 3D printing to reduce the environmental impacts of production. In: ECEEE Industrial Summer Study Proceedings, pp. 213–221 (2014)
10. Gebler, M., Uiterkamp, A.J.S., Visser, C.: A global sustainability perspective on 3D printing technologies. Energy Policy **74**, 158–167 (2014)
11. Li, T.I.N.G.J.I.E., Aspler, J.O.S.E.P.H., Kingsland, A.R.L.E.N.E., Cormier, L.M., Zou, X.U.E.J.U.N.: 3d printing–a review of technologies, markets and opportunities for the forest industry. J. Sci. Technol. For. Prod. Process **5**(2), 30 (2016)
12. Cisneros-López, E.O., Pal, A.K., Rodriguez, A.U., Wu, F., Misra, M., Mielewski, D.F., Kiziltas, A., Mohanty, A.K.: Recycled poly (lactic acid)–based 3D printed sustainable biocomposites: A comparative study with injection molding. Mater. Today Sustain. **7**, 100027 (2020)
13. Benwood, C., Anstey, A., Andrzejewski, J., Misra, M., Mohanty, A.K.: Improving the impact strength and heat resistance of 3D printed models: structure, property, and processing correlationships during fused deposition modeling (FDM) of poly(lactic acid). ACS Omega **3**(4), 4400–4411 (2018)
14. Mohanty, A.K., Vivekanandhan, S., Pin, J.M., Misra, M.: Composites from renewable and sustainable resources: challenges and innovations. Science **362**(6414), 536–542 (2018)

15. Dai, L., Cheng, T., Duan, C., Zhao, W., Zhang, W., Zou, X., Aspler, J., Ni, Y.: 3D printing using plant-derived cellulose and its derivatives: a review. Carbohyd. Polym. **203**, 71–86 (2019)
16. Idrees, M., Jeelani, S., Rangari, V.: Three-dimensional-printed sustainable biochar-recycled PET composites. ACS Sustain. Chem. Eng. **6**(11), 13940–13948 (2018)
17. Huang, S.H., Liu, P., Mokasdar, A., Hou, L.: Additive manufacturing and its societal impact: a literature review. Int. J. Adv. Manuf. Technol. **67**(5–8), 1191–1203 (2013)
18. Ligon, S.C., Liska, R., Stampfl, J., Gurr, M., Mülhaupt, R.: Polymers for 3D printing and customized additive manufacturing. Chem. Rev. **117**(15), 10212–10290 (2017)
19. Nguyen, N.A., Barnes, S.H., Bowland, C.C., Meek, K.M., Littrell, K.C., Keum, J.K., Naskar, A.K.: A path for lignin valorization via additive manufacturing of high-performance sustainable composites with enhanced 3D printability. Sci. Adv. **4**(12), eaat4967 (2018)
20. Mann, G.S., Singh, L.P., Kumar, P., Singh, S.: Green composites: a review of processing technologies and recent applications. J. Thermoplast. Compos. Mater. **33**(8), 1145–1171 (2020)
21. Hung, B.P., Naved, B.A., Nyberg, E.L., Dias, M., Holmes, C.A., Elisseeff, J.H., Dorafshar, A. H., Grayson, W.L.: Three-dimensional printing of bone extracellular matrix for craniofacial regeneration. ACS Biomater. Sci. Eng. **2**(10), 1806–1816 (2016)
22. Zhao, R., Torley, P., Halley, P.J.: Emerging biodegradable materials: starch-and protein-based bio-nanocomposites. J. Mater. Sci. **43**(9), 3058–3071 (2008)
23. Lee, S.G., Xu, X.: Design for the environment: life cycle assessment and sustainable packaging issues. Int. J. Environ. Technol. Manage. **5**(1), 14–41 (2005)
24. Zhang, K., Mohanty, A.K., Misra, M.: Fully biodegradable and biorenewable ternary blends from polylactide, poly(3-hydroxybutyrate-co-hydroxyvalerate) and poly(butylene succinate) with balanced properties. ACS Appl. Mater. Interfaces **4**(6), 3091–3101 (2012)
25. Markstedt, K., Håkansson, K., Toriz, G., Gatenholm, P.: Materials from trees assembled by 3D printing–Wood tissue beyond nature limits. Appl. Mater. Today **15**, 280–285 (2019)
26. Eckert, C., Sharmin, S., Kogel, A., Yu, D., Kins, L., Strijkstra, G.J., Polle, A.: What makes the wood? exploring the molecular mechanisms of xylem acclimation in hardwoods to an ever-changing environment. Forests **10**(4), 358 (2019)
27. Singh, S., Ramakrishna, S., Berto, F.: 3D Printing of polymer composites: a short review. Mater. Des. Process. Commun. **2**(2), e97 (2020)
28. Prakash, K.S., Nancharaih, T., Rao, V.S.: Additive manufacturing techniques in manufacturing-an overview. Mater. Today: Proc. **5**(2), 3873–3882 (2018)
29. Lalegani Dezaki, M., Ariffin, M.K.A.M., Appalanaidoo, D., Wahid, Z., Rage, A.M.: 3D printed object's strength-to-weight ratio analysis for M3 liquid material. Adv. Mater. Process. Technol. 1–15 (2020)
30. Hegab, H.A.: Design for additive manufacturing of composite materials and potential alloys: a review. Manuf. Rev. **3**, 11 (2016)
31. Singh, M., Haverinen, H.M., Dhagat, P., Jabbour, G.E.: Inkjet printing—process and its applications. Adv. Mater. **22**(6), 673–685 (2010)
32. Chakraborty, S., Biswas, M.C.: 3D printing technology of polymer-fiber composites in textile and fashion industry: a potential roadmap of concept to consumer. Compos. Struct. **248**, 112562 (2020)
33. Fu, S.Y., Lauke, B., Mäder, E., Yue, C.Y., Hu, X.J.C.P.A.A.S.: Tensile properties of short-glass-fiber-and short-carbon-fiber-reinforced polypropylene composites. Compos. A Appl. Sci. Manuf. **31**(10), 1117–1125 (2000)
34. Gürcüm, B.H., Börklü, H.R., Sezer, K., Eren, O.: Implementing 3D printed structures as the newest textile form. J. Fashion Technol. Textile Eng. **S4**, 2 (2018). https://doi.org/10.4172/2329-9568.S4-019
35. Kabir, S.F., Mathur, K., Seyam, A.F.M.: A critical review on 3D printed continuous fiber-reinforced composites: history, mechanism, materials and properties. Compos. Struct. **232**, p. 111476 (2020)

36. Partsch, L., Vassiliadis, S., Papageorgas, P.: 3D printed textile fabrics structures. In: The International Istanbul Textile Congress, Istanbul, Turkey (2015)
37. Kamel, S.: Nanotechnology and its applications in lignocellulosic composites, a mini review. Exp. Polym. Lett. **1**(9), 546–575 (2007)
38. Lim, H.W., Cassidy, T.D.: 3D printing technology revolution in future sustainable fashion. In: Sustainability in Textiles and Fashion. Leeds (2014)
39. Lee, J.Y., An, J., Chua, C.K.: Fundamentals and applications of 3D printing for novel materials. Appl. Mater. Today **7**, 120–133 (2017)
40. Chua, C.K., Wong, C.H., Yeong, W.Y.: Standards, Quality Control, and Measurement Sciences in 3D Printing and Additive Manufacturing. Academic Press (2017)
41. Chua, C.K., Yeong, W.Y.: Materials for bioprinting. In: Bioprinting: Principles and Applications, pp. 117–164. World Scientific Publishing Co. Pte. Ltd, Singapore (2014)
42. Lee, J.Y., Tan, W.S., An, J., Chua, C.K., Tang, C.Y., Fane, A.G., Chong, T.H.: The potential to enhance membrane module design with 3D printing technology. J. Membr. Sci. **499**, 480–490 (2016)
43. Lee, J.Y., An, J., Chua, C.K., Fane, A.G., Chong, T.H.: A perspective on 3D printed membrane: direct/indirect fabrication methods via direct laser writing (2016)
44. Ambrosi, A., Moo, J.G.S., Pumera, M.: Helical 3D-printed metal electrodes as custom-shaped 3D platform for electrochemical devices. Adv. Func. Mater. **26**(5), 698–703 (2016)
45. Loo, A.H., Chua, C.K., Pumera, M.: DNA biosensing with 3D printing technology. Analyst **142**(2), 279–283 (2017)
46. Gopinathan, J., Noh, I.: Recent trends in bioinks for 3D printing. Biomater. Res. **22**(1), 11 (2018)
47. Álvarez-Chávez, C.R., Edwards, S., Moure-Eraso, R., Geiser, K.: Sustainability of bio-based plastics: general comparative analysis and recommendations for improvement. J. Clean. Prod. **23**(1), 47–56 (2012)
48. Álvarez-Chávez, C.R., Edwards, S., Moure-Eraso, R., Geiser, K.: Sustainability of bio-based plastics: general comparative analysis. In: 3rd International Workshop on Advances in Cleaner Production, São Paulo, Brazil (2011)
49. Arikan, E.B., Ozsoy, H.D.: A review: investigation of bioplastics. J. Civ. Eng. Arch **9**, 188–192 (2015)
50. Auras, R., Harte, B., Selke, S.: An overview of polylactides as packaging materials. Macromol. Biosci. **4**(9), 835–864 (2004)
51. Auvergne, R., Caillol, S., David, G., Boutevin, B., Pascault, J.P.: Biobased thermosetting epoxy: present and future. Chem. Rev. **114**(2), 1082–1115 (2014)
52. Bechthold, I., Bretz, K., Kabasci, S., Kopitzky, R., Springer, A.: Succinic acid: a new platform chemical for biobased polymers from renewable resources. Chem. Eng. Technol.: Indus. Chem. Plant Equip. Process Eng. Biotechnol. **31**(5), 647–654 (2008)
53. Gandini, A., Lacerda, T.M.: From monomers to polymers from renewable resources: recent advances. Prog. Polym. Sci. **48**, 1–39 (2015)
54. Robert, C., De Montigny, F., Thomas, C.M.: Tandem synthesis of alternating polyesters from renewable resources. Nat. Commun. **2**(1), 1–6 (2011)
55. Bozell, J.J., Petersen, G.R.: Technology development for the production of biobased products from biorefinery carbohydrates—the US department of energy's "Top 10" revisited. Green Chem. **12**(4), 539–554 (2010)
56. Gioia, C., Banella, M.B., Vannini, M., Celli, A., Colonna, M., Caretti, D.: Resorcinol: a potentially bio-based building block for the preparation of sustainable polyesters. Eur. Polymer J. **73**, 38–49 (2015)
57. Schlosser, Š., Blahušiak, M.: Biorefinery for production of chemicals, energy and fuels. Elektroenergetika **4**(2) (2011)
58. Briassoulis, D., Dejean, C., Picuno, P.: Critical review of norms and standards for biodegradable agricultural plastics part II: composting. J. Polym. Environ. **18**(3), 364–383 (2010)

59. Díaz Lantada, A., de Blas Romero, A., Sánchez Isasi, Á., Garrido Bellido, D.: Design and performance assessment of innovative eco-efficient support structures for additive manufacturing by photopolymerization. J. Ind. Ecol. **21**(S1), S179–S190 (2017)
60. Hanssen, O.J.: Environmental impacts of product systems in a life cycle perspective: a survey of five product types based on life cycle assessments studies. J. Clean. Prod. **6**(3–4), 299–311 (1998)
61. Davachi, S.M., Kaffashi, B.: Polylactic acid in medicine. Polym. Plast. Technol. Eng. **54**(9), 944–967 (2015)
62. Drummer, D., Cifuentes-Cuéllar, S., Rietzel, D.: Suitability of PLA/TCP for fused deposition modeling. Rapid Prototyp. J. **18**(6), 500–507 (2012)
63. Hamad, K., Kaseem, M., Yang, H.W., Deri, F., Ko, Y.G.: Properties and medical applications of polylactic acid: a review. Exp. Polym. Lett. **9**(5), 435–455 (2015)
64. Huneault, M.A., Li, H.: Morphology and properties of compatibilized polylactide/ thermoplastic starch blends. Polymer **48**(1), 270–280 (2007)
65. Kreiger, M., Pearce, J.M.: Environmental life cycle analysis of distributed three-dimensional printing and conventional manufacturing of polymer products. ACS Sustain. Chem. Eng. **1**(12), 1511–1519 (2013)
66. Li, J., He, Y., Inoue, Y.: Thermal and mechanical properties of biodegradable blends of poly (L-lactic acid) and lignin. Polym. Int. **52**(6), 949–955 (2003)
67. Mathew, A.P., Oksman, K., Sain, M.: Mechanical properties of biodegradable composites from poly lactic acid (PLA) and microcrystalline cellulose (MCC). J. Appl. Polym. Sci. **97**(5), 2014–2025 (2005)
68. Mekonnen, T., Mussone, P., Khalil, H., Bressler, D.: Progress in bio-based plastics and plasticizing modifications. J. Mater. Chem. A **1**(43), 13379–13398 (2013)
69. Mohanty, A.K., Wibowo, A., Misra, M., Drzal, L.T.: Development of renewable resource–based cellulose acetate bioplastic: effect of process engineering on the performance of cellulosic plastics. Polym. Eng. Sci. **43**(5), 1151–1161 (2003)
70. Sudesh, K., Iwata, T.: Sustainability of biobased and biodegradable plastics. CLEAN–Soil, Air, Water **36**(5–6), 433–442 (2008)
71. Ventola, C.L.: Medical applications for 3D printing: current and projected uses. Pharm. Ther. **39**(10), 704 (2014)
72. Van Wijk, A.J.M., Van Wijk, I.: 3D printing with biomaterials: Towards a sustainable and circular economy. IOS Press (2015)
73. Zhang, Z., Harrison, M.D., Rackemann, D.W., Doherty, W.O., O'Hara, I.M.: Organosolv pretreatment of plant biomass for enhanced enzymatic saccharification. Green Chem. **18**(2), 360–381 (2016)
74. Zhao, B., Chen, G., Liu, Y.U., Hu, K., Wu, R.: Synthesis of lignin base epoxy resin and its characterization. J. Mater. Sci. Lett. **20**(9), 859–862 (2001)
75. Vink, E.T., Rabago, K.R., Glassner, D.A., Gruber, P.R.: Applications of life cycle assessment to NatureWorks™ polylactide (PLA) production. Polym. Degrad. Stab. **80**(3), 403–419 (2003)
76. Rengier, F., Mehndiratta, A., Von Tengg-Kobligk, H., Zechmann, C.M., Unterhinninghofen, R., Kauczor, H.U., Giesel, F.L.: 3D printing based on imaging data: review of medical applications. Int. J. Comput. Assist. Radiol. Surg. **5**(4), 335–341 (2010)
77. ASTM, D.: 638-02a Standard Test Method for Tensile Properties of Plastics, p. 55. ASTM International, West Conshohocken, PA (2002)
78. Standard, A.S.T.M.: D695, 2015, Standard Test Method for Compressive Properties of Rigid Plastics. ASTM International, West Conshohocken, PA (2015)
79. Azimi, P., Zhao, D., Pouzet, C., Crain, N.E., Stephens, B.: Emissions of ultrafine particles and volatile organic compounds from commercially available desktop three-dimensional printers with multiple filaments. Environ. Sci. Technol. **50**(3), 1260–1268 (2016)
80. Bandyopadhyay, A., Bose, S., Das, S.: 3D printing of biomaterials. MRS Bull. **40**(2), 108–115 (2015)

81. Cherykhunthod, W., Seadan, M., Suttiruengwong, S.: Effect of peroxide and chain extender on mechanical properties and morphology of poly(butylene succinate)/poly(lactic acid) blends. IOP Conf. Ser.: Mater. Sci. Eng. **87**, 012073 (2015). https://doi.org/10.1088/1757-899X/87/1/012073

82. Corneillie, S., Smet, M.: PLA architectures: the role of branching. Polym. Chem. **6**(6), 850–867 (2015)

83. Athanasiou, K.A., Niederauer, G.G., Agrawal, C.M.: Sterilization, toxicity, biocompatibility and clinical applications of polylactic acid/polyglycolic acid copolymers. Biomaterials **17**(2), 93–102 (1996)

84. Holowka, E., Bhatia, S.K.: Drug Delivery. Springer-Verlag, New York (2016)

85. Bhatia, S.K.: Traumatic injuries. In: Biomaterials for Clinical Applications, pp. 213–258. Springer, New York, NY (2010)

86. Bhatia, S.K., Bhatia, R.K., Yang, Y.H.: Biosynthesis of polyesters and polyamide building blocks using microbial fermentation and biotransformation. Rev. Environ. Sci. Biotechnol. **15**(4), 639–663 (2016)

87. Mann, G.S., Singh, L.P., Kumar, P., Singh, S., Prakash, C.: On briefing the surface modifications of polylactic acid: a scope for betterment of biomedical structures. J. Thermoplast. Compos. Mater. **34**(7), 977–1005 (2019). https://doi.org/10.1177/0892705719856052

88. Ramakrishna, S., Mayer, J., Wintermantel, E., Leong, K.W.: Biomedical applications of polymer-composite materials: a review. Compos. Sci. Technol. **61**(9), 1189–1224 (2001)

89. Chandramohan, D., Marimuthu, K.: Bio composite materials based on bio polymers and natural fibers-contribution as bone implants. IJAMSAR **1**(1), 09–12 (2011)

90. Santo, V.E., Duarte, A.R.C., Gomes, M.E., Mano, J.F., Reis, R.L.: Hybrid 3D structure of poly(d, l-lactic acid) loaded with chitosan/chondroitin sulfate nanoparticles to be used as carriers for biomacromolecules in tissue engineering. J. Supercrit. Fluids **54**(3), 320–327 (2010)

91. Huang, S.J., Edelman, P.G.: An overview of biodegradable polymers and biodegradation of polymers. In: Degradable Polymers, pp. 18–28. Springer, Dordrecht (1995)

92. Wang, H., Qiu, Z.: Crystallization behaviors of biodegradable poly (l-lactic acid)/graphene oxide nanocomposites from the amorphous state. Thermochim. Acta **526**(1–2), 229–236 (2011)

93. Sandhu, K., Singh, J.P., Singh, S.: Some investigations on the tensile strength of additively manufactured polylactic acid components. In: Advances in Materials Processing, pp. 221–230. Springer, Singapore (2020)

94. Sandhu, K., Singh, S., Prakash, C.: Analysis of angular shrinkage of fused filament fabricated poly-lactic-acid prints and its relationship with other process parameters. IOP Conf. Ser.: Mater. Sci. Eng. **561**(1), 012058 (2019)

95. Lima, L.T., Aurasb, R., Rubinob, M.: Processing technologies for poly (lactic acid). Prog. Polym. Sci. **33**(8), 820–852 (2008)

96. Pandey, J.K., Reddy, K.R., Kumar, A.P., Singh, R.P.: An overview on the degradability of polymer nanocomposites. Polym. Degrad. Stab. **88**(2), 234–250 (2005)

97. Rasal, R.M., Janorkar, A.V., Hirt, D.E.: Poly (lactic acid) modifications. Prog. Polym. Sci. **35**(3), 338–356 (2010)

98. Sandhu, K., Singh, G., Singh, S., Kumar, R., Prakash, C., Ramakrishna, S., Królczyk, G., Pruncu, C.I.: Surface characteristics of machined polystyrene with 3D printed thermoplastic tool. Materials **13**(12), 2729 (2020)

99. Singh, S., Singh, G., Sandhu, K., Prakash, C., Singh, R.: Investigating the optimum parametric setting for MRR of expandable polystyrene machined with 3D printed end mill tool. Mater. Today: Proc. **33**, 1513–1517 (2020)

# Chapter 3
# Is Laser Additive Manufacturing Sustainable?

**C. P. Paul, Sunil Yadav, S. K. Nayak, A. N. Jinoop, and K. S. Bindra**

**Abstract** Laser Additive Manufacturing (LAM) has revolutionized industrial manufacturing by enabling the fabrication of lighter, stronger, complex and customized metallic parts. LAM is attractive mainly due to the various freedoms offered by the technology, like—shape design freedom, material design freedom, logistic freedom and post-processing freedom. LAM technology is considered as a green technology due to two major attributes: minimum material wastage as compared to conventional manufacturing techniques and direct conversion of 3D models to 3D components eliminating intermediate development stages. In addition, LAM possess significant potential to build highly efficient functional components with minimal life cycle impact. Thus, the wider adaptation of LAM is dependent on the sustainability of the technology, which is primarily dependent on economic, environmental and societal effects of LAM. Although LAM paves a way towards sustainability in manufacturing, achieving complete sustainability is challenging due to the use of support structures, higher energy consumption, need for post-processing, etc. Thus, assessing the environmental impact and cost-effectiveness at each stages of the life cycle of LAM built components play a significant role in wider adoption of the technology. This chapter compiles an overview of LAM technology and the various associated processes. It discusses the need for sustainability in manufacturing, the factors governing and challenging the sustainability of LAM technology. A comprehensive comparative assessment of LAM with other conventional manufacturing techniques is compiled and the various theoretical models are discussed in detail. At the end, the suggestions and future directions to improve the sustainability of LAM is deliberated. This chapter is a quick-start for novices to understand this novel technology and a reference document for researchers in the field.

C. P. Paul (✉) · S. Yadav · S. K. Nayak · A. N. Jinoop · K. S. Bindra
Laser Technology Division, Raja Ramanna Centre for Advanced Technology, Indore, Madhya Pradesh 452013, India
e-mail: paulcp@rrcat.gov.in

C. P. Paul · S. Yadav · S. K. Nayak · A. N. Jinoop · K. S. Bindra
Homi Bhabha National Institute, Anushaktinagar, Mumbai, Maharashtra 400094, India

K. Sandhu et al. (eds.), *Sustainability for 3D Printing*, Springer Tracts
in Additive Manufacturing, https://doi.org/10.1007/978-3-030-75235-4_3

**Keywords** Additive manufacturing · Laser additive manufacturing ·
Sustainability · Product life cycle assessment · Environmental impact

## 3.1　Introduction

Industrial metabolism for an economic system involves the conversion of raw
material, energy and labour into goods and services. As per the 2018 annual report
by International Energy Agency (IEA), industrial sector contributes 29% to the total
energy consumption and 42% to the global carbon dioxide emission [1]. In addition, there are rising concerns due to environmental issues, like—increasing carbon
footprint, decreasing non-renewable resources, increasing cost of resources, etc.
Thus, there is a need to develop efficient materials and improved industrial products
to reduce the global energy consumption and emission in industry. The adoption of
the above can result in a strong declination in industrial $CO_2$ emissions due to
improvement in design, material efficiency, material recycling, reuse and substitution. Further, the adoption of efficient design and improved material efficiency can
potentially reduce the global energy usage by 3% per year [1]. Thus, it is necessary
to adopt sustainable manufacturing practices in industries to meet the goals of
sustainable development in manufacturing sector.

Sustainable development has become an integral part of holistic growth of
mankind. Sustainability is about creating an environment in which a suitable
equilibrium can be maintained between social, ecological and economical objectives. In case of industries, sustainability is primarily governed by the manufacturing sector as it involves the conversion of raw material into a final product using
various energy-intensive techniques. In addition, manufacturing contributes to
carbon dioxide emission and primary resource extraction [2]. Thus, assessment of
sustainability of a manufacturing process is essential for sustainable development in
industries. Several research are being carried out globally to categorize framework
for sustainability assessment of various manufacturing process, like—casting,
joining, forming and machining based on environmental, economic and social
impacts at different stages of component life [3]. Environmental impact of a
manufacturing process is broadly assessed in terms of energy consumption, effective material utilization and emission of greenhouse gases. Economic impact is
evaluated based on inventory cost, production cost and production time. Whereas,
social impact assessment is carried out by considering the ease of component usage
or user friendliness of the component, life of the component, customer safety and
environment safety. However, all the assessment terms are inter-related and have
significant impact on the sustainability of manufacturing process. Thus, assessment
is essential during the different stages of the Product Life Cycle (PLC), which
includes designing, fabrication/processing, functional usage/ operational period and
disposal/End-of-Life (EoL) of a component.

One of the techniques to improve the sustainability in industrial manufacturing is
through the deployment of sustainable manufacturing practices in industries,

like—the use of Minimum Quantity Lubrication (MQL) [4] and cryo-cooling technique for machining [5], CRIMSON model for casting [6], etc. In addition, the use of advanced manufacturing processes also paves a way for improving the sustainability of industrial manufacturing. One of the advanced manufacturing processes that can potentially develop highly efficient components with minimal footprint is Additive Manufacturing (AM) [7]. AM possesses potential sustainable features, like—effective material utilization, reduced lead time, capability to fabricate complex and lightweight components, reduced supply chain, etc. [8]. In appending a list of various AM techniques, Metal Additive Manufacturing (MAM) is commonly used in industrial manufacturing to develop customized engineering components. Among the various MAM techniques, Laser Additive Manufacturing (LAM) is widely used for building complex shaped engineering components. LAM leads to effective utilization of material due to addition of material only where it is necessary. In addition, LAM allows for fabrication of lightweight components for engineering and medical applications. LAM also facilitate the part consolidation and development of components with user-defined density. However, few issues such as the need for support structures while fabricating overhang geometry, higher geometrical distortion during fabrication of large size components, higher roughness and higher reflectivity of metals to infrared radiations leads to lower process efficiency. These issues are challenging towards the sustainability of LAM technology. Furthermore, the literature forestalls significant advantage of LAM towards sustainability and its role towards sustainable development [7–9]. Therefore, it is essential to evaluate up to what extent LAM can contribute towards a green and sustainable manufacturing.

This chapter covers the various LAM processes, sustainability assessment of LAM, challenges and scope of LAM towards sustainable development. The sustainability assessment in terms of material utilization, energy consumption, environmental impact and functional advantages of LAM at each stage of product life is explained in detail. In addition, challenges and scope in LAM towards sustainable development are also discussed.

## 3.2  Laser Additive Manufacturing

LAM uses high-power lasers and feedstock material in wire or powder form for layer-by-layer fabrication of engineering components. The wide deployment of LAM is mainly due to the unique advantages of lasers, likeability to process wide varieties of material, generate energy density, be automated etc. In addition, LAM offers various advantages such as shape design freedom, material design freedom, logistics freedom and post-processing freedom [10]. As the fabrication takes place in a layer-by-layer fashion, any complex geometry is converted to simple 2D sketches in each layer, which provides the shape design freedom for LAM. Shape design freedom allows for design of efficient, lightweight and complex components. Material design freedom is provided by LAM mainly due to its ability to add materials only where it is required; add multiple materials in single component; develop compositionally graded components. In addition, the structural variation in material property can be achieved through LAM by controlling microstructural features through user-defined cooling rate during manufacturing. This can be performed by varying the process parameters or using different deposition strategies. The variation in component composition is achieved by varying feedstock compositions or using different feedstock materials to build multi-material or compositionally graded components [11, 12]. LAM offers less resource involvement and reduced physical flow of material leading to lower inventory and logistics cost. In addition, the manufacturing process can be decentralized by LAM as the process requires only a digital model as the input. The file can be sent through cyber systems and components can be developed distantly without direct interaction between manufacturer and customer. LAM also provides post-processing freedom to the user by allowing the user to design the component as per the post-processing requirements. The users have the freedom to tailor the properties, finish and geometry as per the final requirement by using post-processing techniques.

LAM is broadly classified into Laser Powder Bed Fusion (LPBF) and Laser Directed Energy Deposition (LDED) based on material feeding methodology [13] as presented in Fig. 3.1. LPBF system uses laser to selectively melt or sinter regions of a pre-placed bed of powder according to the input design as shown in Fig. 3.2a. After selective melting and building of a layer, a subsequent layer of powder is spread and the process is repeated till the final component is developed in a layer-by-layer fashion. LPBF is broadly classified into Laser Sintering (LS) and, Laser Melting (LM). LS is typically used to build polymer components [10]. However, metallic components are also built using LS during the initial days due to lack of availability of high-power lasers. It uses a multi-material feedstock comprising of metallic powders and binders. Fusion between metal powder and binder is achieved by heating without liquefaction or complete melting of metal powder. However, with the

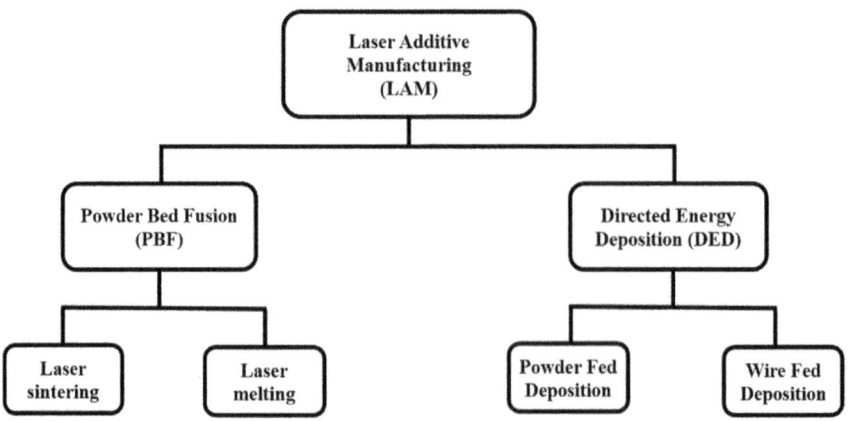

**Fig. 3.1** Classification of LAM

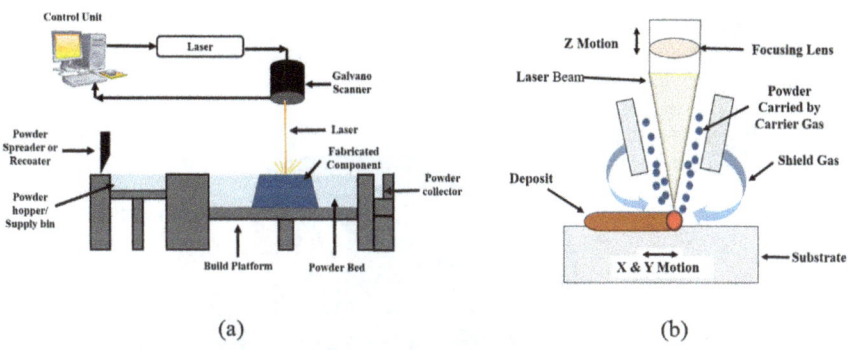

**Fig. 3.2** Schematic of LAM system, **a** powder bed fusion, **b** directed energy deposition

advent of high laser powers, laser energy is enough to completely melt the powder available at the pre-placed bed as per the required design, which leads to the development of entire 3D component, which is known as LM process. In LDED, high-power laser is deployed to generate melt pool on a substrate and feed-stock material (powder or wire form) is dynamically fed onto the melt pool as shown in Fig. 3.2b. The feedstock melts and solidifies after coming in contact with the melt pool and results into the formation of a new deposited layer. This sequential layer-by-layer melt-pool generation, feeding of raw material and melting and solidification leads to development of 3D component as per the input design.

## 3.3  Sustainability of LAM

Since last two decades, Life Cycle Assessment (LCA) and Life Cycle Cost (LCC) have become popular tools to evaluate the impact of product and services on various dimensions of sustainability, i.e. environment, economy and society [14, 15]. The life cycle of any component is broadly divided into two stages: *cradle-to-gate* and *cradle-to-grave* [7, 9]. Cradle-to-gate consists of conversion of raw material into actual product by going through various processing operations. Whereas, cradle-to-grave accounts for product functional life and end of life considering reusability, disposal and recycling aspect of the product at the end of life. In both the cases, designing stage of product life cycle is crucial as it consumes resources in terms of labour and machine. Thus, the current life cycle study of LAM component can be broadly categorized into: early-stage, processing stage, functional stage (operational stage) and end-of-life (EoL). First phase of LAM product life is the early stage or designing phase, which accounts for conversion of concept into design considering feasibility of fabrication. It involves design of a product, i.e. digital model generation, material selection and development of the processing strategy. Additionally, it also involves procurement stages of raw material and inventory control. The second phase of LAM product life is the processing stage.

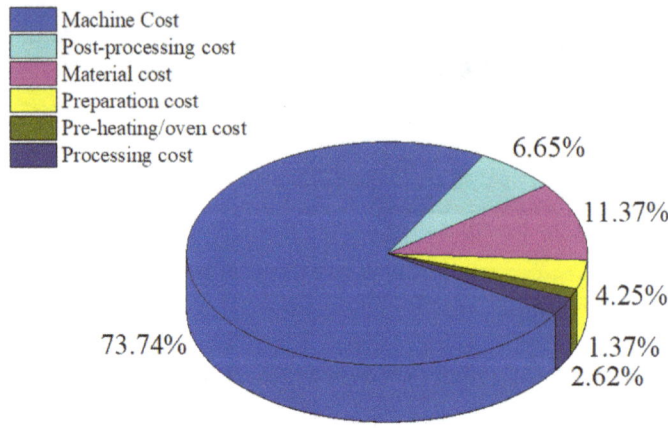

**Fig. 3.3** Cost-breakdown during LAM (Redrawn from [16])

Processing stage includes conversion of raw material into actual product at the expense of energy, material and labour cost. It also involves pre-processing and post-processing activity to achieve desired features or properties in the product. In the third phase, functional stage of a product is considered, which accounts for effectiveness and advantages offered by LAM component during application stage of the product. Finally, the last phase is EoL of product, which explains the disposal, recycling and reusability of LAM product. From Fig. 3.3, it is observed that every stage in LAM consumes significant amount of resources and time, which adds to the production cost with machine cost being the highest followed by material cost and post-processing cost [16]. Post-processing is generally a combination of automatic and manual process. Thus, the cost involved in post-processing is a combination of machine costs and labour costs. Therefore, assessment in terms of material usage, energy usage and labour usage over the entire lifespan is essential.

### 3.3.1  Early Stage of LAM Product Life

Early stage of LAM product life consists of design development and raw material procurement required for the fabrication of the component. Design development involves the development of the digital model based on the inputs from customer, material selection and selection of deposition strategy (like—process parameter selection, build orientation, scanning pattern, etc.). Procurement stage comprises of logistics and inventory of raw material and substrate/build plate material. In addition to the above, additional accessories, like—substrate cooling system, preheating setup, etc. required during fabrication of component, need to be taken care during the early stage of LAM. Every sub-parts in the early stage of LAM product life plays a significant role as they decide the material and energy requirement, material movement and build time. Thus, assessment of every sub-part in the early life of LAM component is very crucial for sustainable manufacturing.

(a) **Assessment of Designing Stage**

LAM offers higher design flexibility as compared to conventional manufacturing techniques, which makes the technique highly suitable for fabricating complex-shaped components. This freedom offered by LAM called "shape design freedom", allows the user to build components by considering the component efficiency instead of manufacturing limitations [10]. In addition, the design freedom offered by LAM also allows the fabrication of lightweight components that permits efficient material utilization and higher efficiency of the component. Lightweight component not only contributes to reduced mass consumption per components, but also offers significant reduction in PLC costs [17]. As per reports by Niaki et al., every kg of reduced mass in aviation component leads to annual fuel savings worth 3000 USD [18]. Moreover, lightweight component also offers less emission during entire life cycle [19].

The design freedom offered by the technology can be put into practice by following the various principles of *Design for Additive Manufacturing (DfAM)*. DfAM offers fabrication advantages, like—inculcation of undercut, variable composition and internal cooling channels. It also provides the freedom for fabrication of complex shapes, such as freeform geometries, bio-inspired designs etc. [20]. DfAM also involves the use of topology optimization algorithms, which aids to redesign a component as per the user requirement. Topology optimization is being used for several years in manufacturing sector. However, with the advent of LAM and DfAM principles, the applications of topology optimization techniques have increased multi-fold times. For example, aerospace components generally prefer lightweight components with higher strength to weight ratio and lower buy-to-fly ratio. With the aid of advanced topology optimization tools, the components can be built for maximum stiffness and minimum weight in order to achieve a higher strength to weight ratio. The major difference between the use of topology optimization tools for LAM and conventional manufacturing is that the use of topology optimization for LAM does not necessities the consideration of the manufacturing constraints during design stage.

One of the design freedoms offered by LAM is the feasibility for fabrication of conformal channels in a component. Conventionally cooling channels are fabricated in straight line causing non-uniform cooling and lower cycle efficiency, especially in mould cavity [21]. Conformal channels offer higher heat transfer and controlled cooling, which helps to improve heat transfer causing higher system efficiency in heat exchanger and injection moulds. For instance, one of the studies indicate that a tool mould fabricated using copper and stainless steel powder consisting of conformal serpentine channels shows a reduction in the cycle time and part distortion by 15% and 37%, respectively [22].

Design freedom offered by LAM also includes customization of a component as per the user design. In addition to the above, LAM combines *customization* and *mass production* via *mass customization*, which allows the fabrication of multiple customized products. This aids to derive the advantages of the mass production and customization in a single build.

Another salient feature of LAM is the part consolidation. LAM possess the capability to transform conventionally assembled component built using functionally different parts into a single component. This results in a superior component with reduced parts, assembly and production cost [23]. In addition, reduced assembly eliminates quality problems arising due to assembly.

The design stage cost is majorly contributed by the labour cost, utility cost of the system and software charges [17]. One of the advantages offered by LAM as compared to other manufacturing processes is that the variation of cost is independent of the complexity of component as opposed to conventional manufacturing [24] as shown in Fig. 3.4. In addition, the ability of LAM to incorporate product customization at any stage aids for quick response to customer needs and customer satisfaction. Due to the widespread availability of CAD software and LAM systems, the consumers are moving away from being passive to being actively involved in the production process in order to produce customized products and hence becoming *prosumers* within the global manufacturing community [7]. Thus, design for LAM components promote sustainability in terms of low cost and social advantages in term of customer-centric process.

(b) **Assessment of Procurement Stage**

Quality and consistency of components fabricated by LAM is highly dependent on the quality of feedstock material used for processing. Thus, procurement of optimum quality feedstock material is crucial for defect-free component fabrication [25]. Procurement stage of LAM involves purchase of raw material and inventory management. Powder based LAM uses metal powder of few microns to hundreds of microns (typically 5–150 µm). Fine powder in the range of 5–50 µm is generally

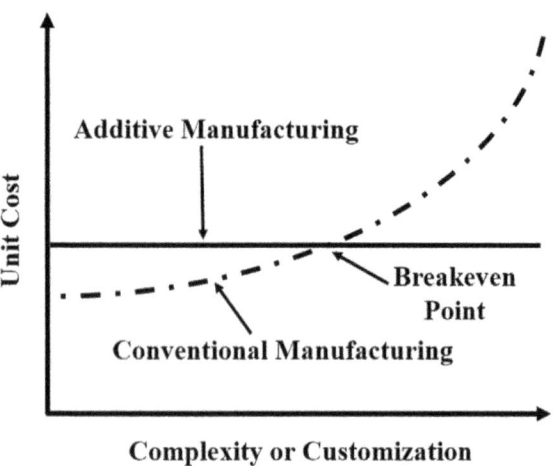

**Fig. 3.4** Product unit cost versus complexity of the component (Redrawn from [24])

used for LPBF systems and the maximum size of powder governs the layer thickness, which indirectly determines the minimum feature size in the build direction. LDED systems typically uses powders of size ranging from 50 to 150 µm. The minimum value of powder size is limited by the ability of the powder to flow to the melt pool during deposition. If the powder size is too large, it can result in in-complete melting and larger feature size [26, 27]. These powders are generally produced by atomization techniques, especially gas or plasma atomization techniques. Wire is used as a feedstock material during DED based LAM and they are typically fabricated by drawing process. Generally, the cost of wire is lower than the cost of powder for the same material.

Layer-by-layer near-net-shaped fabrication using LAM can lead to lower buy-to-fly ratio ($\sim 1$) as compared to conventional manufacturing, which causes reduced raw material requirement [28]. For example: an aircraft component fabricated by subtractive routes (milling, cutting and turning) incurred buy-to-fly ratio between 8 and 30, whereas the buy-to-fly ratio is close to 1.5 for the same component built using LAM process [19, 29, 30]. Thus, volume of raw material required during LAM is lower than conventional manufacturing techniques. These aspects potentially causes environment-friendly supply chain due to reduced volume of material and shorter transportation chain [7]. In addition, low volume requirement also ensures low inventory cost. In one of Fraunhofer ILT demonstration, 70% material saving and 30% time cost saving is achieved with LDED [7]. However, the production of metal powder is cumbersome as it involves significant amount of resources and energy and causes equivalent environmental impacts, which leads to higher cost as compared to raw material production for conventional manufacturing processes [25].

Economic analysis of procurement stage mainly involves feedstock material cost, storage/ inventory cost and transportation cost. Most commonly used feedstock material in LAM is powder and as mentioned earlier the unit price of powder is typically higher than solid or bulk material. However, with LAM, the amount of material required to build the component is lower than conventional manufacturing due to low buy-to-fly ratio, which results in cost savings. During LAM, the total volume of powder used will have three major components: volume of powder used for building the component, volume of powder used for support structure and extra volume of powder removed in the build chamber after fabrication. Thus, volume of built component is typically lower than the volume of powder added to the powder hopper/powder supply bin for fabrication. As a consequence, there is a scope to reuse the unused powder available in the build chamber/ hopper/ powder supply bin. Procurement cost can be written as:

$$C_{\text{proc}} = n_{\text{vp}} * \left(C_{\text{vp}} + C_{\text{transp}} + C_{\text{invent}}\right) \tag{3.1}$$

where $C_{\text{proc}}$ is the net procurement cost, $C_{\text{vp}}$ is the per-unit cost of virgin powder, $C_{\text{transp}}$ is the per-unit transportation cost, $C_{\text{invent}}$ is the per-unit inventory cost and

$n_{vp}$ is the number of units of virgin powder. Net procurement cost depends on the number of units of virgin powder used for fabrication. It is worthy to note that material cost contributes to 10–11% of overall production cost [16, 31].

Thus, it can be concluded that the design freedom offered by LAM provides greater potential towards sustainable manufacturing. Reduced number of components and assembly, efficient complex design, lower buy-to-fly ratio, direct involvement of customer and reduced inventory and transportation in early stage of LAM component life have direct impact on environmental, economic and societal aspect of sustainability.

### 3.3.2 Processing Stage of LAM Product Life

Processing stage of LAM comes under the "cradle-to-gate" stage of a component and it consists of building components in a layer-by layer-fashion as per the design input. Processing stage involves maximum utilization of energy, material and manpower. Thus, this stage possesses significant potential, which affects the utilization, economic and environmental aspect of LAM. Therefore, assessment of production or development stage of LAM product life is crucial. Once the detailed design of component is released, fabrication starts as early as possible. This leads to reduced time delay between design and fabrication stage, which results in significant cost savings.

LAM offers fabrication of product without using any physical tool, i.e. tool-free fabrication. Conventional fabrication of metallic components involves tools and moulds requiring time and cost on changeover in design. Thus, tool-free fabrication features endow LAM to produce tailored component in a sustainable approach [32]. In addition, tool-free fabrication promotes on-demand production culture that minimizes inventory and logistics [33].

Several researchers are investigating a comparison between additive and subtractive manufacturing techniques on resource consumption and environmental aspects [17, 30, 34, 35]. LAM promotes sustainability in terms of maximum utilization of material during processing stage as the processing volume of raw material in LAM is lower due to near-net-shape fabrication in a layer-by-layer fashion. In addition, limited tool accessibility offered by conventional machining makes it difficult or impossible during fabrication of complex-shaped components [36]. For material removal operation, assessment is based on the volume of chip produced and environmental impact of the process, which is expressed in terms of "K ratio" (mass of raw material to mass of final component) and "R ratio" (environmental impact of LAM strategy to environmental impact of conventional strategy), respectively [34]. K ratio represents material removal volume during machining whereas, R ratio represents environmental impact. For higher value of K ratio (>5), chip volume is higher leading to higher consumption of energy and cutting fluid causing significant impact on environment. Whereas, for lower chip volume, i.e. for lower value of $K$ ratio, energy consumption and environmental impact reduces by using conventional machining process. Thus, for conventional machining, small and medium chip volume shows minimum energy consumption. Whereas, for higher chip volume, i.e. for higher value of $K$ ratio, energy consumption and processing time is higher. If the value of $K$ ratio is lower ($\leq 3$) and geometry of component is simple, the use of conventional manufacturing processes for production will be beneficial in terms of environmental impact with lower value of R ratio. However, for fabrication of complex-shaped components, LAM have higher advantages due to reduced material consumption and environmental impact [34]. Thus, Specific Energy Consumption (SEC) and emissions during fabrication of complex component is lower for LAM as compared to the material removal processes [19].

LAM by its very nature does not result in the production of chips or scraps and therefore results in very less material wastage and related environmental impact as opposed to conventional manufacturing techniques such as machining. In conventional machining (e.g. CNC machining), the ratio of material waste to the final part can reach up to 19:1 [37]. Also, due to minimal wastage of materials without scarifying the product quality, LAM can be a potential lean manufacturing technique. Since LAM is capable of producing final product in one go, there is limited exposure to hazardous conditions and therefore little hazardous waste. As mentioned in previous sections, LAM can build components with lower buy-to-fly ratio as compared to conventional manufacturing and it is one of the widely considered topic in aviation sector. For instance, the problem of higher buy-to-fly ratio is addressed under European Union (EU) FP7 MERLIN project using LAM. Under this project, DED is used to fabricate bladed disks to be used in aero-engines. The process avoided the generation of swarf, which cannot be recycled economically. With DED material savings and time savings of approximately 60% and 30%, respectively are achieved [7].

In case of formative manufacturing, like—casting, moulding, forging and stamping, mould, die and tool are together called as "tooling". These are essential for metal forming and shaping operations in structural, aerospace, automotive, medical and electronic industries. Fabrication of these tools are complex, time intensive and expensive as they involve customized materials, complex designs, skilled labour requirement and complex fabrication procedures [38]. For example, the cost of die for a land gear in aeroplane incurred 91% of total cost of landing gear [20]. In addition, functional life of these tools are limited due to corrosive environment, repeated thermal loading and intensive production cycles [39]. As LAM starts directly from digital model data, the development of moulds and tooling using LAM will be a viable option for quick replacement of tools and moulds. This application of LAM is widely known as "Rapid Tooling". Several studies have reported the fabrication of tooling's using LDED and LPBF system [39–42]. One of the case study indicates the deployment of LDED reduces lead time by 35% [38]. In addition to the quick replacement, LAM can aid in improving the sustainability of formative manufacturing by reducing the cycle time and improving the life to the mould by inserting conformal cooling channels and providing compositionally

**Table 3.1** Comparison of machining, casting and LAM technology based on various manufacturing aspects

| Manufacturing aspects | Machining or material removal technique | Casting/metal forming technique | LAM |
|---|---|---|---|
| Volume of component | Preferred for low chip volume producing design ($K$ ratio <3) | Independent of volume | Depends on build volume and workstation limits of the LAM machine |
| Material of the component | Hardness of material affects the energy consumption and coolant requirement | Energy consumption depends on the thermophysical properties of the material | Energy consumption depends on the thermophysical properties of the material |
| Tooling requirement | Tool, tool insert and coolant is required | Moulds and die are required | Tool-free fabrication |
| Time | Short lead time | Relatively higher lead time | Shorter lead time |
| Complexity of the component | Lower | Limited, mould and die design add to cost | Flexible designs possible and no additional cost requirement |
| Inventory size | High | High | Comparatively low |

graded tooling's with enhanced efficiency [40–42]. Table 3.1 presents the qualitative comparison of machining, casting and LAM technology based on various manufacturing aspects.

Sustainability during processing stage can be well explained using energy consumption and material utilization. Sustainability in manufacturing could be significantly improved by reducing the energy consumption during the processing phase. There have been consistent efforts to improve the efficiency of the manufacturing process during the processing stage. Some of the efforts include energy monitoring, analysis and reporting [43]. The processing stage of LAM deals with the energy consumption and material utilization during the fabrication of a desired part. The amount and the fraction of energy being utilized by different machine parts and ancillary parts during the process also varies among different LAM processes. Feng et al. modelled energy utilization during LS based LAM process as presented in Eq. 3.2 [17];

$$E_{processing} = E_{laser} + E_{heating} + E_{maintain} + E_{delivery} + E_{control} + E_{aid} \qquad (3.2)$$

where $E_{processing}$, $E_{laser}$, $E_{heating}$, $E_{maintain}$, $E_{delivery}$, $E_{control}$, $E_{aid}$ are energy utilized in complete processing, energy utilized by laser head, energy used in bringing the chamber/substrate to a particular temperature, energy used by heating element to maintain a particular temperature in the build chamber, energy utilized in the material feed, energy used for controlling the mechanical station and energy utilized by other elements and aiding functions, respectively.

Energy consumed to sinter unit volume of powder ($e_{laser}$) is defined by Eq. 3.3.

$$e_{laser} = \frac{\omega I_{avg}}{v_{laser}} \tag{3.3}$$

Hence, the total energy required to melt the part can be given by Eq. 3.4

$$E_{laser} = e_{laser}.V_B \tag{3.4}$$

where, $\omega$ is the material absorptivity, $I_{avg}$ is the average laser intensity, $E_{laser}$ is the laser rated power and $V_B$ is the build part's effective build volume.

The laser beam creates a parabolic melt-pool/sintered region in the cross section of the substrate. For such a parabola with gap distance $h_{laser}$ and layer thickness $d_{laser}$, the area is given by Eq. 3.5.

$$A_s = 2 * \frac{h_{laser} d_{laser}}{3} \tag{3.5}$$

Therefore, the total laser energy to sinter the whole part is given by Eq. 3.6

$$E_{laser} = \frac{3.\omega.P_{laser}}{2.v_{laser} h_{laser} d_{laser}} . V_B \tag{3.6}$$

Also, $E_{heating}$ and $E_{maintain}$ are given by Eqs. 3.7 and 3.8 respectively.

$$E_{heating} = \left( M_{powder} C_{powder} + M_{support} C_{support} \right) \Delta T \tag{3.7}$$

$$E_{maintain} = \int_0^{t_{finish}} S_{oven}(T - T_{ext}).k \tag{3.8}$$

where, $M_{powder}$ is the mass of the powder; $M_{support}$ is the mass of the support structure; $C_{powder}$ is the specific heat capacity of the powder; $C_{support}$ is the specific heat capacity of the support structure material; $S_{oven}$ is the surface area of oven; $k$ is the total heat transfer coefficient of oven wall; $T$ is the inside temperature of the oven; $T_{ext}$ is the external ambient temperature; $\Delta T$ is the temperature difference between $T$ and $T_{ext}$.

The energy consumed by the powder delivery system can be defined by Eq. 3.9.

$$E_{delivery} = P_{delivery} \times t_d \tag{3.9}$$

where, $P_{delivery}$ and $t_d$ refers to the average power of the powder delivery system, powder delivery time, respectively.

Another aspect involved in the sustainability of LAM process is the degree of material utilization. The degree of material utilization and material wastage in LAM process is highly different from those in conventional processes, like—machining and forming. The degree of material utilization also varies between different LAM processes, while building the same part. Apart from the fabrication process, the degree of material utilization also depends upon the complexity of the desired geometry, requirement of support structures and deposition strategies. As discussed earlier, there is an increasing research interest towards the deployment of topology optimization for LAM built components. However, these structures often require significant amount of supports. The need for supports increases the material wastage and reduces the material utilization efficiency and increases the fabrication and post-processing costs of LAM.

There are some recent works reporting the energy and material utilization in LAM. Mirzendehdel et al. suggested a topologically optimized model using a novel topological sensitivity approach for constraining support structure volume during design optimization yielding designs with reduced support structures, which is suitable for deposition based AM processes [44]. Strano et al. suggested an approach for optimizing the orientation of the build part and support structures to build cellular structure. It involved providing more supports in the regions of higher weight, which in turn resulted in significant material and cost savings and thereby increasing the sustainability and efficiency of the AM process [45].

Another aspect of the cost of LAM system is determined by the speed of the fabrication of the part. It depends upon the maximum number of parts that can be fabricated at a time by the machine which, in turn depends upon the build volume and degree of democratisation of the process. According to Yoon et al., energy consumed in the heater system, roller drives, stepper motor and piston control, laser system and miscellaneous functions for LS based PBF process is 37%, 16%, 26%, 16%, 5%, respectively. It is observed from the literature that SEC decreased with an increase in the machine deposition productivity. It also revealed that energy consumed in producing a single part is more than that in building multiple parts simultaneously [31]. Yoon et al. compared the maximum SEC for different commercially available LPBF systems using Selective Laser Sintering (SLS)—a process similar to LS based PBF and Direct Metal Laser Sintering (DMLS) and Selective Laser Melting (SLM)—processes similar to LM based PBF. SLS, DMLS and SLM systems were compared from 3D Systems, EOS GmbH and MTT (former name of SLM Solutions), respectively [46]. It was observed that SLS, DMLS and SLM systems under study have maximum SEC of 66.02 kWh/kg, 94.17 kWh/kg and 163.33 kWh/kg, respectively. Sreenivasan et al. carried out an energy assessment in a 3D Systems Vanguard HiQ + HS SLS system and an average operating power of 20 kW larger than EOS based SLS and MTT based SLM system was observed. Out of the total power consumption, powder feed and part bed heaters consumed 40% of the power. The feed and build piston stepper motors consumed 40% of the energy, while the laser system consumed 16% of the energy. It was concluded that energy consumption can be reduced by better thermal management and eliminating preheating of the powder [9]. Baumers et al. observed that LS like every other novel processes with build rate less than 0.1 kg/h are likely to consume energy more than 100 MJ per kilogram of the material processed [31]. In addition, Jackson et al. observed that SEC of powder-based LDED system is higher than wire-based LDED system by 22% and 24% for developing a sub-sized tensile bar and plate tensile bar, respectively. However, the powder-based LDED consumed 3% less energy for fabricating sheet sized tensile samples as compared to wire-based LDED process [47].

LAM processing stage includes utilization of material, energy and labour for specified duration of time. Thus, the utilization of material, energy and labour contribute to the production cost in both direct and indirect form. The direct cost includes cost of the material and the energy usage, whereas, indirect cost includes machine cost, labour cost, consumable cost and overhead cost. Labour cost includes design loading cost, supervision cost, machine build-setup preparation cost and post-processing cost. The labour cost is generally considered as a fixed cost [19, 48, 49]. In addition, LAM includes risk cost, which accounts for build failure that might be caused due to process instability [50]. However, risk cost arising due to material of the component can be reduced by optimizing processing parameters and strategy. Breakdown of the cost reveals that indirect cost contributes to 36.53% of the overall production cost followed by labour cost, risk cost, material cost and energy cost (excluding energy consumption of auxiliary accessories) of 26.69% 26%, 10.22% and 0.29%, respectively [31]. However, overall production cost depends on the build quantity and the material of the component [51]. The build quantity governs the material usage and the material of the component governs the energy usage. Thus, both the build quantity and the material of component have significant impact on the total cost. In addition, fabrication cost for building a single component will be higher if the complete build volume is not utilized. As the build volume utilization increases, the fabrication cost decreases during LAM. Capacity utilization of the build volume can be increased by fabricating a larger number of components by effectivity utilizing the space in the build plate/substrate. However, the maximum number of components that can be built in a build platform is limited by the size of the build platform, size of the component and minimum gap to be maintained between the components for defect-free fabrication [16, 52].

One of the factors that contribute to the production cost is the fabrication time or build time. As the build time increases, the cost associated with the energy usage and the cost associated with the labour increases, leading to a higher production cost. Higher build speed results in higher production rate causing reduction in overall production cost. If different LAM processes are compared, fabrication time is higher in LPBF than LDED typically due to smaller layer thickness. However, the build rate of LAM is lower than conventional route of manufacturing. For example, the deposition rate of DMLS is typically 37.58 g/h whereas for conventional fabrication (machining and casting) processing rate in the order of 1,00,000 g/h [50]. Further, cost associated during production stage includes both the fixed cost (machine and labour cost) $C_{fixed}$ and the building cost [16]. Building cost is the direct cost arising from the material and the energy consumption over the building time. Production cost can be expressed as $C_{production}$ (refer Eq. 3.10)

$$C_{production} = C_{fixed} + (building\ time * building\ rate) \tag{3.10}$$

For LDED system build rate is evaluated by considering the cost of the material and the energy consumed during the fabrication. It includes the cost of shielding gas, carrier gas, laser energy and material utilized during the fabrication stage. Whereas, for the LPBF system, the cost associated with the inert atmosphere, the energy consumed by the spreader or roller for powder spreading, material and laser energy need to be considered in the building cost.

Baumer et al. reported a method to express the cost of LAM process by using a "r" factor (refer Eq. 3.11), which shows the fraction of the cost of energy and raw material used with respect to the total cost incurred during the development of a component. Typical "r" value for LDED is 0.12, which indicates 0.12 unit is actually utilized in material and energy conversion into product out of 1 unit used during processing phase [50].

$$r = \frac{(unit\ of\ raw\ material\ consumed * unit\ price\ of\ raw\ material) + (unit\ price\ of\ energy * unit\ of\ energy\ consumed)}{overall\ cost}$$

$$\tag{3.11}$$

Thus, it can be concluded that due to the tool-free fabrication nature of LAM and its efficient utilization of material and energy, LAM possess significant potential for fabricating complex-shaped and multi-material component sustainably. Despite these, the issues associated with LAM such as slower build rate, higher roughness, anisotropy in mechanical properties and higher production costs need to be addressed carefully for affordable and sustainable manufacturing.

### 3.3.3 Functional Life

The functional life of a component deals with the duration for which the component is in application/service. The life of a component is being continuously affected by its service loading conditions and the service environment. The life of the component can generally be improved by adopting optimal geometrical designs or adding a protective layer or a combination of both. Achieving the above objectives are relatively easy through LAM as compared to conventional techniques due to higher degree of shape design freedom and the ability for adding materials at specific locations, which can improve the components' functionality. LAM can also be used to generate geometrically optimized components to achieve desired structural and material integrity with minimal adverse effect on the environment. These advantages of LAM are being used mainly by the aviation sector to generate components that are design efficient, material efficient and environment friendly.

In the aviation sector, the efficient lightweight designs are built using LAM, which results in material savings. This weight reduction also leads to saving of fuels, which in turn results of reduction of expenses. Red et al. observed that AM can lead to savings of aluminium, titanium and nickel by 4050 tons/year, 7600 tons/year and 8110 tons/year, respectively by 2050. Also, it is estimated that 1 kg weight reduction in the fan blades could lead to reduction of 0.5 kg in rotor and engine weight [19]. With LAM, a large number of lightweight structures with naturally inspired design can be built according to the concepts of biomimicry [7]. Examples of novel lightweight structures that can be created by LAM are porous mesh arrays and open cellular foams. These structures can enhance strength, stiffness, energy efficiency and corrosion resistance when they are incorporated into the core of the components [7]. These concepts are used in projects such as belt buckles in aeroplanes redesigned for reduced weight, heat sinks for improved airflow and higher thermal efficiency; heat exchangers with constrained geometries that have improved efficiency [7]. LAM can also be used to redesign components in order to be fabricated in an integrated fashion which reduces the need for the coupling hardware. These improved designs can improve the system efficiency considerably. In aviation sector, part consolidation can reduce the fuel emissions by about 75% [53]. The Leading Edge Aviation Propulsion (LEAP) engine launched by General Electric (GE) in 2016 includes LAM built full nozzles that are five times stronger for aiding durability and for providing optimized geometry for fuel flow and more efficient combustion. This new design is a single component as opposed to the previous design, which had an assembly of 20 separate parts. This led to a substantial weight reduction of 25% with respect to the existing nozzles. In addition, part consolidation can lead to reduced possibility of stress concentration zones resulting in reduced possibility of failure and in turn improved life of the part. Part consolidation can also reduce the contact wear between the parts, which also contributes to the improved life of the part.

In bio-medical sector, medical implants can be made lightweight with the help of LAM without compromising the strength. The implants are made lightweight by providing porous features inside a complex shaped component using LAM that was previously impossible by conventional techniques. In automotive sector, manufacturing of spare parts is another market segment that can be potentially transformed by LAM. Spare parts that are hard to obtain can be remotely manufactured using LAM. This helps in reducing the transportation cost and inventory cost, which aids to reduce the associated effect on the environment. These spare parts can also help in restoration and preservation of antique objects.

One of the exciting applications of LAM is fabrication of components with functionally graded materials that can be achieved by providing gradient in material composition in the component. The grading allows reduction in stress concentration yielding improved life in the loading conditions and at the same time saving excess material from use by judicious addition of multiple materials for achieving maximum efficiency. Similar procedure can also be used to provide protective layers over a component. This process is being termed as laser cladding.

As mentioned previously, LAM can also be used to build moulds with conformal cooling channels. Conformal channels offer controlled cooling cycle and the graded composition offers reduced failure at the interface caused by repeated thermal loading [39, 54]. One of the examples of using multi-material tooling is the deposition of highly conductive Cu on SS mould, which enhanced the cycle time by 25% [38, 55]. The principle of using conformal cooling channels was used by Finnish based company Salcomp, a world leader in the production of electrical plugs and power supplies to build injection moulding device with redesigned vent structure using direct metal laser sintering (DMLS) for mobile phones. The cooling time reduced from 14 to 8 s was observed with the use of redesigned structure, which enabled the production of 56,000 extra units each month. Additionally, improvement in quality was observed along with a reduction in rejection rates from 2 to 1.4% [7].

Thus, it can be concluded that by considering the costs, energy and emissions, it is observed that the potential for sustainability occur over the entire functional life of LAM products. Products built by LAM are now finding increasing use for different applications. Some of these applications are in the field of aerospace, medical, tooling sectors etc. Owing to different freedoms with regards to geometry and materials, LAM can be used to build highly complex components in terms of shape or tailored material properties that can provide efficient material usage, less energy consumption and very less $CO_2$ emissions with respect to the conventionally built products. These qualities have shown that LAM built components can be sustainable over their entire functional life.

### 3.3.4  End of Life (EoL)

In order to deal with the depletion of natural resources and environmental impacts due to industrial processes, many environmental policies and regulations, for example, ELV (End-of-Life Vehicle directives) and WEEE (Waste Electrical and Electronic Equipment directives) as presented by Gehin et al. are put into place to limit the amount of landfilled waste [56]. They are responsible for making industries produce their products in a cleaner and greener manner by efficient use of energy and resources. When a product has lived its utility, then it is said to retire or reach its EoL. The increase in the number of EoL products will impact the burden on the landfill activities, which affects the environment considerably. Therefore, it is imperative to develop efficient strategies to recycle EoL products, while simultaneously considering its implications on the environment [34].

As discussed in one of the previous sections, LAM has the advantage of reduced raw material usage. During LAM, reusability of powder should be taken care for making LAM product acceptable and economically viable for industries. In general, feedstock powder can be completely melted, partially melted or heated during the processing. Significant volume of powder is left un-melted that can be separated from the partially melted and condensate powder, which can be reused again as a feedstock. Several studies have indicated that powder reuse for LAM does not result in loss of mechanical properties and build quality [57–59]. For instance, studies report that reusing the powders of stainless steel (SS) 17-4PH, SS 316L and IN718 up to 11, 12 and 14 times, respectively did not affect the mechanical properties significantly [57, 58, 60]. However, it should be noted that excess reusability of powder can hamper fabricated component's property and thus the number of reuse cycles are crucial to determine the final component quality. Thus, it is advisable to use a mixer of virgin and used powder for fabrication in early stage of recycling. However, after a number of reuse cycles, powder can be used for the fabrication of non-critical part of scientific study [25, 61]. Therefore, assessment of powder cost involves both cost of virgin as well as cost of used/ recycled powder. It is observed that the implementation of the powder reusing results in a material utilization efficiency up to 95% [57].

Though, powder reusing enhances the material utilization during processing stage, it incurred additional cost by adding labour and machine cost to the production cost. On the other hand, powder reusing is a manual process and the powder gets exposed to atmosphere during handling. Moisture trapping inside the powder and oxide formation during reusing stage is another limitation of powder reusing [60]. Thus, drying and preheating is necessary to remove the trapped moisture and ensure uniform powder flow. However, the cost associated with the labour and preheating is not significant as compared to the cost saved by using reused powder. Thus, powder reusing paves a way towards improved utilization of feedstock materials leading to affordable and green manufacturing.

Even after the product retires it may hold some residual value. To increase the residual value several EoL options are used including remanufacturing, recycling, reusing, repairing and disposal [17]. LAM potentially has the ability to recover the value embedded in the waste [7]. Traditionally, the LAM products that could be treated as waste could be upcycled to manufacture luxury products using LAM [62]. Recycled LAM waste can also be used to produce metallic powders. One of the case-study on the recycling of LAM components is reported by Cacace et al. The authors attempted to recycle LAM components to generate powders for LAM. The generated powders are further used to build tensile specimens using LPBF and LDED [63]. The built samples show tensile strength values of $653 \pm 1.79$ MPa and $675 \pm 2$ MPa for AISI 316L samples using LPBF and LDED, respectively. This indicates that the components built using recycled powders possesses good processability and high mechanical properties [63]. Similarly, support materials can also be recycled after use [53].

As discussed earlier, buy-to-fly ratio close to 1 can be achieved with LAM with recycling and reusing waste products. Literature indicates that up to 40% of the raw material wastage can be saved, while 95% of them can be reused. This also helps in decreasing the total primary energy supply (TPES) requirements for industrial waste processing or material recycling [53]. The complexity of the processes during recycling increases with the diversity of the materials entering the recycling system.

## 3.4  Challenges and Scope

Even though LAM has several advantages, there are few challenges that need to be considered for wide adoption of the technology. The challenges can be broadly classified as: *processing challenges* and *economic challenges*. Processing challenges such as support structure for overhang geometries, lower build rate, poor finish, higher reflection of metals when processed using industrial infrared lasers, anisotropy in mechanical properties, instability of processes at higher build height, etc. should be overcome for sustainable processing through LAM. Economic challenges primarily include the higher machine costs, feedstock costs, processing costs etc. This section will discuss some of the challenges of LAM affecting sustainability:

(a)  **Support Structure**

One of the biggest challenges while fabricating complex structure is the likelihood of encountering overhang geometry. The fabrication of the overhang geometry invites possibility of the loss in the dimensional accuracy and may result in the failure of the component. Thus, support structures are needed to build overhang features when there is no underlying solid layer to support. Support structures are sacrificial extra material required to support overhang parts during fabrication. For metal AM, it is observed that building overhang structures with overhang angle less than 45° (with respect to horizontal plane) without support structure is difficult [64]. The use of support structure increases the material, energy and labour cost. Support structures add up material and consume energy during the fabrication stage. After fabrication, removal of these support structures is necessary and is achieved by post-cleaning operations. For example, in metal AM, post-cleaning operation cost is approximately 8% of total cost of product [65]. The surface area that is in direct contact with component affects the surface finish of the component. Thus, minimization of support structure is essential to reduce consumption of material and energy over the build time for sustainable manufacturing. In addition, optimum design and material-efficient support structures are required for efficient and sustainable LAM process. Therefore, optimizing the quantity of support structures is one of the growing interests among the research community, as complete elimination of the overhang structures is not possible. Various methods being used for optimizing quantity of support structures are selection of build orientation, designing material-efficient support structures, applying the principles of DfAM, etc.

(b)  **Slower Build Rate**

Slower build rate is another challenge of LAM, which significantly affects the sustainability of the process. Typically, build rate during LAM is governed by the processing parameters (laser power, scan speed and material feed rate or powder layer thickness). As mentioned before smaller build rate in LAM is mainly due to smaller layer thickness in LPBF and slower motion of mechanical system (nozzle head or substrate) in LDED. Slower build rate not only add direct cost to the product, but also causes delay in the product delivery [50, 66]. Generally, slower production results in higher material footprint, carbon footprint and energy footprint by 2.2 times, 2.9 times and 2.2 times, respectively as compared to quick production [67]. The build rate can be improved in LAM by using high-power lasers and higher material feed rate, which increases the material utilization efficiency and build rate.

(c) **Delamination and Cracking**

Delamination and cracking are the common issues arising due to large difference in the thermophysical properties of substrate and build material. They can also occur between different deposited layers due to difference in thermophysical properties arising due to sharp thermal gradient during the process. If the induced residual stress during LAM is higher than ultimate strength of the build material, delamination/cracking can occur. Delamination incurs extra material and energy cost that leads to extra cost in production. It adds to risk cost which is 26% of the total product cost [31]. However, preheating and optimizing process parameters can reduce the temperature gradient and control the cooling rate resulting in reduced or negligible probability for delamination/ cracking [68, 69].

(d) **Higher Reflectivity to Infrared Lasers**

Fibre lasers are the most commonly used lasers for advanced LAM systems. The major advantages of fibre laser that attracts its wide usage are transmission flexibility, capability of achieving higher laser powers, compact designs etc. Fibre laser generates radiation in the infrared regime. However, the lower absorption coefficient of metals in the infrared regime results in inefficient utilization of laser energy during fabrication. The typical reflectivity values for most commonly used metals and alloys are: 60% for SS, >99% for Cu, >96% for Al [70]. However, the above issue can be solved by preheating of the substrate or by using modified powder compositions. Researchers are also investigating the use of lasers with wavelength in the visible regime (blue and green) for processing of highly reflective materials, like—copper, aluminium etc.

(e) **Dimensional Inaccuracy and Surface Roughness**

Though LAM is capable of fabricating near-net-shaped components, the components built using LAM suffers from high surface roughness and dimensional inaccuracy. Poor surface finish and inaccurate dimension not only causes energy and material wastage during the processing stage, but also incurs labour and tooling cost for finishing operations. Lower surface roughness and higher dimensional accuracy can be achieved by controlling the process parameters or process conditions. In addition, post-processing operations can also be used to improve the surface finish and dimensional accuracy of the built components. Some of them include: surface finishing, abrasion finishing, grinding, chemical polishing, milling, etc. Some of the LAM systems uses a hybrid approach by combining additive techniques with conventional milling to build components with higher precision.

(f) **Anisotropy in Mechanical Properties**

Higher cooling rate and continuous thermal cycling during processing leads to anisotropy and non-homogeneous microstructure on LAM built component. This results in anisotropy in mechanical properties, which makes the components unsuitable for several engineering applications. The anisotropy in mechanical property is generally resolved using heat-treatment cycles. Heat-treatment aids to achieve uniform microstructure and isotropic mechanical properties. On the other side, it results in loss of energy and time. However, the anisotropy can be solved to a greater extend by achieving equiaxed microstructure during fabrication. Studies have suggested that this can be achieved by controlling process parameters during fabrication or by using modified LAM processes, like—deployment of ultrasonic waves during LAM [71, 72].

## 3.5 Summary

Sustainable development has become an integral part of holistic growth of mankind. Rising concerns about sustainability can be seen due to environmental issues, increasing carbon footprint, decreasing non-renewable resources, increasing resources cost, rise in approach of customer towards eco-friendly product etc. Therefore, sustainability of a manufacturing process is an essential part towards a sustainable development. AM possess potential sustainable features, like—effective utilization of material, reduced lead time, capability to fabricate complex and lightweight component, reduced supply chain, etc. Out of various AM techniques, Laser Additive Manufacturing (LAM) is one of the advanced fabrication techniques deployed for fabrication of metallic component. LAM is broadly classified into LPBF and LDED based on the material feeding mechanism.

This chapter presented an overview of different sustainability aspects for LAM system. It is observed that LAM possess significant potential and promising future for sustainable development by offering design, material, logistics and post-processing freedom. LCA reveals that effective utilization of material, reduced emission, capability for fabricating complex part, reduced inventory and decentralized market, paves way towards sustainable manufacturing. However, issues such as higher setup and raw material cost, lower build rate and anisotropic mechanical properties in the build part needs to be addressed carefully to further improve the sustainability of LAM process.

**Acknowledgements** S. Yadav, S. K. Nayak and A. N. Jinoop acknowledge the financial support of Raja Ramanna Centre for Advanced Technology (RRCAT), Department of Atomic Energy, Government of India and Homi Bhabha National Institute, Mumbai. The authors express their sincere gratitude to Mr. Debashis Das, Director RRCAT for his constant support and encouragement. Thanks are due to Mr. S. V. Nakhe, Director—Laser Group for constant encouragement in this evolving program at RRCAT. The authors are also thankful to our colleagues at Laser Additive Manufacturing Laboratory for their support.

# References

1. World Energy Outlook 2018—Analysis—IEA, (n.d.). https://www.iea.org/reports/world-energy-outlook-2018. Accessed 20 Sept. 2020
2. Priarone, P.C., Ingarao, G.: Towards criteria for sustainable process selection: on the modelling of pure subtractive versus additive/subtractive integrated manufacturing approaches. J. Clean. Prod. **144**, 57–68 (2017). https://doi.org/10.1016/j.jclepro.2016.12.165
3. Saad, M.H., Nazzal, M.A., Darras, B.M.: A general framework for sustainability assessment of manufacturing processes. Ecol. Indic. **97**, 211–224 (2019). https://doi.org/10.1016/j.ecolind.2018.09.062
4. Sharma, V.S., Singh, G., Sorby, K.: A review on minimum quantity lubrication for machining processes. Mater. Manuf. Process. **30**, 935–953 (2015). https://doi.org/10.1080/10426914.2014.994759
5. Bayraktar, S.: Cryogenic cooling-based sustainable machining. In: High Speed Machining, pp. 223–241. Elsevier (2020). https://doi.org/10.1016/b978-0-12-815020-7.00008-4
6. Dai, X., Jolly, M.: Potential energy savings by application of the novel CRIMSON aluminium casting process. Appl. Energy. **89**, 111–116 (2012). https://doi.org/10.1016/j.apenergy.2010.12.029
7. Ford, S., Despeisse, M.: Additive manufacturing and sustainability: an exploratory study of the advantages and challenges. J. Clean. Prod. **137**, 1573–1587 (2016). https://doi.org/10.1016/j.jclepro.2016.04.150
8. Peng, T., Kellens, K., Tang, R., Chen, C., Chen, G.: Sustainability of additives manufacturing: an overview on its energy demand and environmental impact. Addit. Manuf. **21**, 694–704 (2018). https://doi.org/10.1016/j.addma.2018.04.022
9. Sreenivasan, R., Goel, A., Bourell, D.L.: Sustainability issues in laser-based additive manufacturing. Phys. Procedia **5**, 81–90 (2010). https://doi.org/10.1016/j.phpro.2010.08.124
10. Paul, C.P., Jinoop, A.N., Bindra, K.S.: Metal additive manufacturing using lasers. In: Singh, R., Davim, J.P. (eds.) Additive Manufacturing: Applications and Innovations, pp. 37–94. CRC Press, Florida (2018)
11. Yadav, S., Jinoop, A.N., Sinha, N., Paul, C.P., Bindra, K.S.: Parametric investigation and characterization of laser directed energy deposited copper-nickel graded layers. Int. J. Adv. Manuf. Technol. **108**, 3779–3791 (2020). https://doi.org/10.1007/s00170-020-05644-9
12. Banait, S.M., Paul, C.P., Jinoop, A.N., Kumar, H., Pawade, R.S., Bindra, K.S.: Experimental investigation on laser directed energy deposition of functionally graded layers of Ni–Cr–B–Si and SS316L. Opt. Laser Technol. **121**, 105787 (2020). https://doi.org/10.1016/j.optlastec.2019.105787
13. Nayak, S.K., Mishra, S.K., Paul, C.P., Jinoop, A.N., Bindra, K.S.: Effect of energy density on laser powder bed fusion built single tracks and thin wall structures with 100 μm preplaced powder layer thickness. Opt. Laser Technol. **125**, 106016 (2020). https://doi.org/10.1016/j.optlastec.2019.106016
14. ISO—ISO 14040:2006—Environmental Management—Life Cycle Assessment—Principles and Framework, (n.d.). https://www.iso.org/standard/37456.html. Accessed 20 Sept, 2020
15. ISO—ISO 14044:2006—Environmental Management—Life Cycle Assessment—Requirements and Guidelines, (n.d.). https://www.iso.org/standard/38498.html. Accessed 20 Sept. 2020
16. Lindemann, C., Jahnke, U., Moi, M., Koch, R.: Analyzing product lifecycle costs for a better understanding of cost drivers in additive manufacturing (2012)
17. Ma, J., Harstvedt, J.D., Dunaway, D., Bian, L., Jaradat, R.: An exploratory investigation of additively manufactured product life cycle sustainability assessment. J. Clean. Prod. **192**, 55–70 (2018). https://doi.org/10.1016/j.jclepro.2018.04.249
18. Niaki, M.K., Torabi, S.A., Nonino, F.: Why manufacturers adopt additive manufacturing technologies: the role of sustainability. J. Clean. Prod. **222**, 381–392 (2019). https://doi.org/10.1016/j.jclepro.2019.03.019

19. Huang, R., Riddle, M., Graziano, D., Warren, J., Das, S., Nimbalkar, S., Cresko, J., Masanet, E.: Energy and emissions saving potential of additive manufacturing: the case of lightweight aircraft components. J. Clean. Prod. **135**, 1559–1570 (2016). https://doi.org/10.1016/j.jclepro. 2015.04.109

20. Atzeni, E., Salmi, A.: Economics of additive manufacturing for end-usable metal parts. Int. J. Adv. Manuf. Technol. **62**, 1147–1155 (2012). https://doi.org/10.1007/s00170-011-3878-1

21. Shinde, M.S., Ashtankar, K.M.: Additive manufacturing–assisted conformal cooling channels in mold manufacturing processes. Adv. Mech. Eng. **9**, 168781401769976 (2017). https://doi. org/10.1177/1687814017699764

22. Sachs, E., Allen, S., Guo, H., Banos, J., Cima, M., Serdy, J., Brancazio, D.: Progress on Tooling by 3D Printing; Conformal Cooling, Dimensional Control, Surface Finish and Hardness, (n.d.)

23. Boothroyd, G.: Product design for manufacture and assembly. Comput. Des. **26**, 505–520 (1994). https://doi.org/10.1016/0010-4485(94)90082-5

24. Pinkerton, A.J.: [INVITED] Lasers in additive manufacturing. Opt. Laser Technol. **78**, 25–32 (2016). https://doi.org/10.1016/j.optlastec.2015.09.025

25. Dawes, J., Bowerman, R., Trepleton, R.: Introduction to the additive manufacturing powder metallurgy supply chain. Johnson Matthey Technol. Rev. **59**, 243–256 (2015). https://doi.org/ 10.1595/205651315X688686

26. Spierings, A., Levy, G.: Comparison of density of stainless steel 316 L parts produced with selective laser melting using different powder grades (2009)

27. Liu, B., Wildman, R., Tuck, C., Ashcroft, I., Hague, R.: Investigation the effect of particle size distribution on processing parameters optimisation in selective laser melting process (2011)

28. Cozmei, C., Caloian, F.: Additive manufacturing flickering at the beginning of existence. Procedia Econ. Financ. **3**, 457–462 (2012). https://doi.org/10.1016/s2212-5671(12)00180-3

29. Barz, A., Buer, T., Haasis, H.D.: A study on the effects of additive manufacturing on the structure of supply networks. IFAC-PapersOnLine **49**, 72–77 (2016). https://doi.org/10.1016/ j.ifacol.2016.03.013

30. Green Manufacturing: Degrees of Perfection (n.d.): http://green-manufacturing.blogspot.com/ 2010/07/degrees-of-perfection.html. Accessed 20 Sept. 2020

31. Baumers, M., Beltrametti, L., Gasparre, A., Hague, R.: Informing additive manufacturing technology adoption: total cost and the impact of capacity utilisation. Int. J. Prod. Res. **55**, 6957–6970 (2017). https://doi.org/10.1080/00207543.2017.1334978

32. Niaki, M.K., Nonino, F.: The management of additive manufacturing—enhancing business value (2018).https://doi.org/10.1007/978-3-319-56309-1

33. Berman, B.: 3-D printing: the new industrial revolution. Bus. Horiz. **55**, 155–162 (2012). https://doi.org/10.1016/j.bushor.2011.11.003

34. Le, V.T., Paris, H., Mandil, G.: Environmental impact assessment of an innovative strategy based on an additive and subtractive manufacturing combination. J. Clean. Prod. **164**, 508–523 (2017). https://doi.org/10.1016/j.jclepro.2017.06.204

35. Ingarao, G.: Manufacturing strategies for efficiency in energy and resources use: the role of metal shaping processes. J. Clean. Prod. **142**, 2872–2886 (2017). https://doi.org/10.1016/j. jclepro.2016.10.182

36. Zhu, Z., Dhokia, V., Newman, S.T.: The development of a novel process planning algorithm for an unconstrained hybrid manufacturing process. J. Manuf. Process. **15**, 404–413 (2013). https://doi.org/10.1016/j.jmapro.2013.06.006

37. Liu, Z., Jiang, Q., Zhang, Y., Li, T., Zhang, H.C.: Sustainability of 3D printing: a critical review and recommendations. In: ASME 2016 11th International Manufacturing Science and Engineering Conference MSEC 2016, American Society of Mechanical Engineers (2016). https://doi.org/10.1115/MSEC2016-8618

38. Morrow, W.R., Qi, H., Kim, I., Mazumder, J., Skerlos, S.J.: Environmental aspects of laser-based and conventional tool and die manufacturing. J. Clean. Prod. **15**, 932–943 (2007). https://doi.org/10.1016/j.jclepro.2005.11.030

39. Bragg, L.M., Miller, M.E., Koplan, S., Rogowsky, R.A., Simpson, V.: Tools, Dies, and Industrial Molds: Competitive Conditions in the United States and Selected Foreign Markets, (n.d.)
40. Zeng, G.H., Song, T., Dai, Y.H., Tang, H.P., Yan, M.: 3D printed breathable mould steel: small micrometer-sized, interconnected pores by creatively introducing foaming agent to additive manufacturing. Mater. Des. **169**, 107693 (2019).https://doi.org/10.1016/j.matdes. 2019.107693
41. Lušić, M., Barabanov, A., Morina, D., Feuerstein, F., Hornfeck, R.: Towards zero waste in additive manufacturing: A case study investigating one pressurised rapid tooling mould to ensure resource efficiency. Procedia CIRP **37**, 54–58 (2015). https://doi.org/10.1016/j.procir. 2015.08.022
42. Renkó, J.B., Kemény, D.M., Nyiro, J., Kovács, D.: Comparison of cooling simulations of injection moulding tools created with cutting machining and additive manufacturing. Mater. Today Proc. **12**, 462–469 (2019). https://doi.org/10.1016/j.matpr.2019.03.150
43. Giret, A., Trentesaux, D., Prabhu, V.: Sustainability in manufacturing operations scheduling: a state of the art review. J. Manuf. Syst. **37**, 126–140 (2015). https://doi.org/10.1016/j.jmsy. 2015.08.002
44. Mirzendehdel, A.M., Suresh, K.: Support structure constrained topology optimization for additive manufacturing. CAD Comput. Aided Des. **81**, 1–13 (2016). https://doi.org/10.1016/j. cad.2016.08.006
45. Strano, G., Hao, L., Everson, R.M., Evans, K.E.: A new approach to the design and optimisation of support structures in additive manufacturing. Int. J. Adv. Manuf. Technol. **66**, 1247–1254 (2013). https://doi.org/10.1007/s00170-012-4403-x
46. Yoon, H.S., Lee, J.Y., Kim, H.S., Kim, M.S., Kim, E.S., Shin, Y.J., Chu, W.S., Ahn, S.H.: A comparison of energy consumption in bulk forming, subtractive, and additive processes: review and case study. Int. J. Precis. Eng. Manuf. Green Technol. 1(3), 261–279 (2014). https://doi.org/10.1007/s40684-014-0033-0
47. Jackson, M.A., Van Asten, A., Morrow, J.D., Min, S., Pfefferkorn, F.E.: Energy consumption model for additive-subtractive manufacturing processes with case study. Int. J. Precis. Eng. Manuf. Green Technol. 5(4), 459–466 (2018). https://doi.org/10.1007/s40684-018-0049-y
48. Rickenbacher, L., Spierings, A., Wegener, K.: An integrated cost-model for selective laser melting (SLM). Rapid Prototyp. J. **19**, 208–214 (2013). https://doi.org/10.1108/ 13552541311312201
49. Piili, H., Happonen, A., Väistö, T., Venkataramanan, V., Partanen, J., Salminen, A.: Cost estimation of laser additive manufacturing of stainless steel. Phys. Procedia **78**, 388–396 (2015). https://doi.org/10.1016/j.phpro.2015.11.053
50. Baumers, M., Dickens, P., Tuck, C., Hague, R.: The cost of additive manufacturing: machine productivity, economies of scale and technology-push. Technol. Forecast. Soc. Change. **102**, 193–201 (2016). https://doi.org/10.1016/j.techfore.2015.02.015
51. Ruffo, M., Hague, R.: Cost estimation for rapid manufacturing ' simultaneous production of mixed components using laser sintering. Proc. Inst. Mech. Eng. Part B J. Eng. Manuf. **221**, 1585–1591 (2007). https://doi.org/10.1243/09544054JEM894
52. Rapid Manufacturing: An Industrial Revolution for the Digital Age, Wiley, (n.d.). https:// www.wiley.com/en-in/Rapid+Manufacturing%3A+An+Industrial+Revolution+for+the+Digital +Age-p-9780470016138. Accessed 20 Sept. 2020
53. Gebler, M., Schoot Uiterkamp, A.J.M., Visser, C.: A global sustainability perspective on 3D printing technologies. Energy Policy **74**, 158–167 (2014). https://doi.org/10.1016/j.enpol. 2014.08.033
54. Noecker, F.F., DuPont, J.N.: Functionally graded copper—steel using laser engineered net shaping ᵀᴹprocess. In: ICALEO 2002—21st International Congress on Applications of Laser Electro-Optics, Congress Proceedings, Laser Institute of America, p. 185430 (2002). https:// doi.org/10.2351/1.5066217
55. Injection Molds and Molding: A Practical Manual—J.B. Dym—Google Books, (n.d.). https:// books.google.co.in/books?hl=en&lr=&id=dVcMzY_c2skC&oi=fnd&pg=PR9&dq=injection

+molding+handbook.+Dordrecht,+The+Netherlands:+Kluwer+Academic%3B+2000&ots=7FjR7EjNLM&sig=u4s9MvD-m7Zeh86OdqoisAKve5M#v=onepage&q&f=false. Accessed 20 Sept. 2020

56. Gehin, A., Zwolinski, P., Brissaud, D.: A tool to implement sustainable end-of-life strategies in the product development phase. J. Clean. Prod. **16**, 566–576 (2008). https://doi.org/10.1016/j.jclepro.2007.02.012

57. Ardila, L.C., Garciandia, F., González-Díaz, J.B., Álvarez, P., Echeverria, A., Petite, M.M., Deffley, R., Ochoa, J.: Effect of IN718 recycled powder reuse on properties of parts manufactured by means of selective laser melting. Phys. Procedia **56**, 99–107 (2014). https://doi.org/10.1016/j.phpro.2014.08.152

58. Jacob, G., Brown, C., Donmez, A., Watson, S., Slotwinski, J.: NIST advanced manufacturing series 100–6. Effects of powder recycling on stainless steel powder and built material properties in metal powder bed fusion processes, (n.d.). https://doi.org/10.6028/NIST.AMS.100-6

59. Santecchia, E., Spigarelli, S., Cabibbo, M.: Material reuse in laser powder bed fusion: side effects of the laser—metal powder interaction. Metals (Basel) **10**, 341 (2020). https://doi.org/10.3390/met10030341

60. Gorji, N.E., O'Connor, R., Mussatto, A., Snelgrove, M., González, P.G.M., Brabazon, D.: Recyclability of stainless steel (316 L) powder within the additive manufacturing process. Materialia **8**, 100489 (2019).https://doi.org/10.1016/j.mtla.2019.100489

61. Clayton, J., Deffley, R.: Optimising metal powders for additive manufacturing. Met. Powder Rep. **69**, 14–17 (2014). https://doi.org/10.1016/S0026-6657(14)70223-1

62. Braungart, M., W.M.-D. la cuna a la, undefined 2013, Cradle to cradle, (n.d.)

63. Cacace, S., Furlan, V., Sorci, R., Semeraro, Q., Boccadoro, M.: Using recycled material to produce gas-atomized metal powders for additive manufacturing processes. J. Clean. Prod. **268**, 122218 (2020).https://doi.org/10.1016/j.jclepro.2020.122218

64. Thomas, D., R.B.-P. of T.C. Technologies, undefined 2008, Identifying the geometric constraints and process specific challenges of selective laser melting, (n.d.)

65. Thomas, D.S., Gilbert, S.: Costs and cost effectiveness of additive manufacturing: a literature review and discussion. Disaster resilient communities view project measuring the level of under-reporting of wildland fires view project (2016).https://doi.org/10.6028/NIST.SP.1176

66. Gutowski, T.G., Branham, M.S., Dahmus, J.B., Jones, A.J., Thiriez, A., Sekulic, D.P.: Thermodynamic analysis of resources used in manufacturing processes. Environ. Sci. Technol. **43**, 1584–1590 (2009). https://doi.org/10.1021/es8016655

67. Teubler, J., Weber, S., Suski, P., Peschke, I., Liedtke, C.: Critical evaluation of the material characteristics and environmental potential of laser beam melting processes for the additive manufacturing of metallic components. J. Clean. Prod. **237**, 117775 (2019).https://doi.org/10.1016/j.jclepro.2019.117775

68. Alimardani, M., Toyserkani, E., Huissoon, J.P., Paul, C.P.: On the delamination and crack formation in a thin wall fabricated using laser solid freeform fabrication process: an experimental-numerical investigation. Opt. Lasers Eng. **47**, 1160–1168 (2009). https://doi.org/10.1016/j.optlaseng.2009.06.010

69. Mukherjee, T., Zhang, W., DebRoy, T.: An improved prediction of residual stresses and distortion in additive manufacturing. Comput. Mater. Sci. **126**, 360–372 (2017). https://doi.org/10.1016/j.commatsci.2016.10.003

70. Siva Prasad, H., Brueckner, F., Volpp, J., Kaplan, A.F.H.: Laser metal deposition of copper on diverse metals using green laser sources. Int. J. Adv. Manuf. Technol. **107**(3–4), 1559–1568 (2020). https://doi.org/10.1007/s00170-020-05117-z.

71. Yadav, S., Paul, C.P., Jinoop, A.N., Rai, A.K., Bindra, K.S.: Laser directed energy deposition based additive manufacturing of copper: process development and material characterizations. J. Manuf. Process. **58**, 984–997 (2020). https://doi.org/10.1016/j.jmapro.2020.09.008

72. Todaro, C.J., Easton, M.A., Qiu, D., Zhang, D., Bermingham, M.J., Lui, E.W., Brandt, M., StJohn, D.H., Qian, M.: Grain structure control during metal 3D printing by high-intensity ultrasound. Nat. Commun. **11**, 1–9 (2020). https://doi.org/10.1038/s41467-019-13874-z

# Chapter 4
# Application of Multi-attribute Decision Making Methods for Fused Deposition Modelling

Sagar U. Sapkal and Pritam H. Warule

**Abstract** Fused Deposition Modelling (FDM) is accountably more used 3D printing process because this process has more flexibility to build complex parts. FDM is layered manufacturing process and is highly affected by a number of working variables. Many research based on the optimum combination of working variables for 3D printing process by aid of conventional and recent optimization techniques. The parametric optimization methods are effectively used to understand conflicting nature of different attributes to increase features of part quality and dimensional accuracy. The main branch of optimization methods is Multi-Criteria Decision Making (MCDM), which was further divided into Multi-Attribute Decision Making (MADM) and Multi-Objective Decision Making (MODM). MADM methods are easy to understand and apply, which includes Simple Additive Weighting (SAW), Weighted Product Method (WPM), Preference Ranking Organization Method for Enrichment Evaluations (PROMETHEE), Analytic Hierarchy Process (AHP), Technique for Order Preference by Similarity to Ideal Solution (TOPSIS), Grey Relational Analysis (GRA), etc. This study mainly includes parametric optimization of FDM working variables with the help of MCDM methods. Multi-attribute methods are applied to experimental data of FDM process parameters. The problem for optimization taken on the I-optimality criteria applied to FDM process. The mathematical model showing nonlinear relation of working variables and geometric precision is considered while applying MADM methods. The input parameters taken for optimization are layer thickness, air gap, raster angle, build orientation, road width and number of contours, including response variables as percentage change in length, width and thickness. After application of MADM methods to the selected alternatives and attributes, the methods under consideration have shown different rankings. The ranking is determined on the basis of percentage error between best results shown by earlier researcher and the result shown by each selected alternative. Final ranks for these results are determined by the combination of ranking given by percentage error method and applied MADM method and the concluded best rank can be utilized for further applications. Best ranking gives the productive concatenation of working

S. U. Sapkal (✉) · P. H. Warule
Walchand College of Engineering, Sangli, Maharashtra, India

© The Author(s), under exclusive license to Springer Nature Switzerland AG 2022        55
K. Sandhu et al. (eds.), *Sustainability for 3D Printing*, Springer Tracts
in Additive Manufacturing, https://doi.org/10.1007/978-3-030-75235-4_4

variables which are responsible for dimensional accuracy of FDM of 3D printed part. Also, the comparative evaluation of MADM methods under consideration is carried out and it is found that PROMETHEE method shows the best and more accurate results for this 3D printing process.

**Keywords** Fused deposition modelling · Multi-attribute decision making · PROMETHEE method

## 4.1 Introduction

In upcoming era, because of globalization, the market screenplay has become more competitive and mutable for the manufacturing sectors [1]. The ASTM calls additive manufacturing (AM) as the "process of layer upon layer joining materials to make objects from 3D modelling, resisted by subtractive production techniques by" [2]. The main benefits of the RP process requires no part-specific tools and is fully automated despite the recent introduction of this technique, it has been used in multiple areas, like mechanical part fabrication, construction, routine tools and medical applications [3]. Reduction of product life cycle period is a main problem in industries to secure market competitiveness that is why the conventional method to develop the product was replaced by rapid manufacturing techniques such as Rapid Prototyping [4].

Even though AM involve a different process but FDM is common in most cases [2]. FDM is a highly building rapid prototyping (RP) process because, it is capable to make workable structures with complicated geometrical parts in manageable construction time [4]. Using FDM technique 3D printed tool also used for machining [5, 6]. Response measures of RP methods are mainly impacted due to its performance variable. Because of this, multiple research has focused on finding the optimal combination of performance variable of RP processes by conventional and advanced optimization techniques listed as Teaching–learning-based optimization (TLBO) algorithm, Non-dominated Sorting TLBO (NSTLBO) algorithm [1], Genetic Algorithm (GA), Neural Network, Levenberg–Marquardt optimality theory [3], Taguchi, Differential evolution (DE) [4], Preference Ranking Organization Method for Enrichment Evaluations (PROMETHEE) [7], Jaya Algorithm [8], Petri net modelling [9], Heuristic search, Integrated Computer-Aided Manufacturing function (IDEFO) modelling, Fault tree analysis, Weights-gradient methods, Group method for data handling (GMDH), etc. Optimization means "minimize the effort required, maximize the desired benefits". MODM have many of choices of decision variables, the best of which follows the user's parametric limits and decide the preferences (ranking order). MADM is discrete, with a finite number of options, i.e. the finite number predefined choices. The MADM technique shows how attributes are to be processed to get the desired choice. The MODM method includes Non-Dominated Sort GA (NSGA-II), TLBO Algorithm, Jaya Algorithm and so on.

## 4.2   Literature Review

In view of the application of soft-computing techniques and the Multiple Criteria Decision Making (MCDM) methods of AM following literature has been reviewed. Rao and Rai [1], used TLBO algorithm for a single variable and NSTLBO algorithm for multivariable function with processing material as acrylonitrile butadiene styrene (ABS). The process parameters considered are construction time, product quality and dimensional precision, manufacturing costs and the saved energy during the method. The first case study includes compressive strength obtained from TLBO algorithm is higher than the value obtained by using the QPSO algorithm. And the second case study determined the optimum value of sliding wear with the help of TLBO algorithm, or population size is less than obtained using the QPSO algorithm for sliding wear.

Noriega et al. [2], done dimensional precision enrichment of square cross-section fused deposition modelling (FDM) components using ANN with ABS as a component material.

Rong-Ji et al. [3], worked on optimizing the performance variables for selective laser sintering with the help of GA and neural network inputs are layer thickness, layer capacity, contour width, scanning speed, cycle time, working surrounding temperature and scanning mode. Support materials used are polycarbonate, nylon, composite, wax, etc. for these weights—gradient methods and Levenberg–Marquardt optimality theory were used. Neural network's collected shrinkage ratio gives information about SLS's shrinkage compensation.

Rayegani and Onwubolu [4], got predicting and optimizing FDM process parameters using Data Handling Group Method (GMDH) and Differential Evolution (DE) with ABS as a part material. Parameters taken for optimization are part orientation, raster angle, raster width and air gap show results that improvement in the tensile strength is mainly due to the negative air gap and smaller raster width. Keeping part orientation as zero; raster angle is increased with increases in tensile strength. Rao and Patel [7], made choice of cutting fluid by the aid of PROMETHEE method for a machining application and chooses the Rapid prototyping process. PROMETHEE approach is helped by considering any range of quantitative and qualitative selection techniques for making a decision in the manufacturing environment. While AHP is an aid to calculate the weightages applied to each attribute with a consistency check. Rao [8], developed a normal and latest optimization algorithm namely as Jaya algorithm to solve restricted and unconstrained optimisation problems. Jaya algorithm excludes the control parameters specific to the algorithms; only common control variables are required to arrive at the best solution and the worst solution should be avoided. Yim and Lee [9], did scheduling of Flexible Manufacturing Systems (FMS) using Petri Net Modelling and comparative (heuristic) search methods. Input data for search method are no. of a machine, no. of jobs result into, search methods involve limited memory and computation time; future analysis of weight in heuristic functions. Adinarayana et al. [10], applied Genetic Algorithm (GA) for the three objective functions with less variance to determine the best value of variables with different weight factors. Using a genetic algorithm with L27 Taguchi orthogonal array, the input variables are speed, feed and depth of cut taken to optimize surface features, material removal rate and power consumption when turning EN24 with PVD coated alloy steel.

Mahdavinejad et al. [11], optimizes milling parameters of an ANN. Input parameters for experiments are cutting speed, feed and depth of cut with workpiece material as Ti-6Al-4 V Titanium alloy. Total neural network programming performed by using Matlab software resulting as small feed rates gives best surface roughness.

Chen et al. [12], used improved TLBO algorithm to get the overall optimization by taking inputs as a maximal evolutionary generation, variables size, population size, appropriate solution to the problem, initial population, probability, etc. show results accuracy of the improved TLBO much higher than the remaining techniques and improvement in the contrast of TLBO using a hybridisation of method is possible.

Mahmood et al. [13], analyzed the performance of Flexible Manufacturing System (FMS) with the help of Integrated Computer-Aided Manufacturing function (IDEFO) modelling and Fault tree analysis (FTA). Material used as a steel C35 (HB = 150) with inputs taken is Cycle time, release blank and collect part. Results show a decrease in throughput time and enhancement in productivity and utilization; because of the renovation, the FMS is stable.

Chaudhrya and Khanb [14], studied survey carried out: an analysis of the techniques of flexible job shop scheduling. The review shows that parameters like minimum makespan, total workload of machines is the most repetitive parameters optimized by using most used optimization techniques like the Genetic Algorithm, Hybridization of genetic algorithm, etc.

Venkataramanaish [15], used not only the NP-Hard technique for addressing the missing operation but also simulated annealing (SA) algorithm for minimization of a weighted sum of makespan in cellular manufacturing system (CMS) by a heuristic approach. For optimization, two sets are used with input parameters as makespan, flow time, ideal time, scheduling cost, etc. The obtained result shows that the dispatching rule is better than the algorithm when the percentage of missing operations is more than half.

As per the literature review, it was observed that most of the work is done in the Additive Manufacturing process optimization field. Algorithms and various kinds of optimization techniques are used to optimize process parameters of the FDM process. In that TLBO, TOPSIS, GRA, etc. are used earlier and SAW, WPM, AHP, PROMETHE and Jaya are not applied for more number of manufacturing processes. To apply such optimization techniques the converging rate and accuracy should be as high as possible. To overcome this, the software platforms help to get more accuracy and a high converging rate in less operational time. Hence to apply Metaheuristic for optimization of FDM process parameters this study is focused on comparison and to execute the applicability of software platforms for optimization methods.

## 4.3   Optimization Methods

### 4.3.1   Overview of Optimization Methods

From the study, it has been viewed that, various optimizations techniques, algorithms are available to get the best results, selection of combinations of optimum alternatives, etc. The basic MADM methods are SAW, WPM, AHP, PROMETHEE, TOPSIS, GRA, etc. While Jaya algorithm and TLBO algorithm is the MODM methods has given the best optimum solution compared with other available techniques. Not all available MADM or MODM methods applicable to optimize all types of data, it mainly depends on the decision-maker's requirements. SAW is also known as weighted sum methods. It is simple and still extensive in use. The main thing is to assign the relative score to each attribute as per decision-maker's requirements (i.e. important attribute has to assign the more score) from that final weights are calculated that are applied (sum of multiplication of each normalized value by its relatives' calculated weight) to each identical units of normalized data. Normalization is the main step before the application of calculated weight; it is processed to convert the beneficial and non-beneficial attribute as 1 and remaining attributes made as less than 1. WPM is nearly similar as that of SAW only differs in the product of each normalized value raised to the power of the relative weight of the corresponding attribute. AHP method is always aid to solve the problems of decision making. AHP earlier applied for the selection of AM processes as well as for layered manufacturing techniques. This method is having provision to check the consistency of calculated relative weight (consistency ratio < 0.1). For each criterion, PROMETHEE conducts a pairwise comparison of alternatives to evaluate partial binary relationships that denotes preference power [7] of one parameter over others. A speciality of this method is that it does not require normalized data and both qualitative, quantitative data also used. The final ranking is decided based on the decreasing range of Net Flow (individual difference of the sum of all elements of $i$th row to sum of all elements of $i$th column) value and it is also used for selection of rapid prototyping processes. TOPSIS approach works on a criteria that the selected alternative must have minimal Euclidean distance [16]. This method aid in multi-criteria optimization of part builds orientation through a combined meta-modelling. Grey Relational Analysis (GRA) also used in the FDM process to decide dimensional accuracy as well as for optimization of the process parameter.

### 4.3.2   Multiple Attribute Decision Making (MADM) Methods

This includes four main parts combined in decision table (Decision matrix).

i. Alternatives
ii. Attributes
iii. Weight of each attribute, or relative importance
iv. Alternative performance metrics about attributes.

$A_i$ (for $i$ = 1, 2, 3 ... $N$)—Alternatives
$B_j$ (for $j$ = 1, 2, 3 ... $M$)—Attributes
$w_j$ (for $j$ = 1, 2, 3, ..., $M$)—Weights of attributes and
$m_{ij}$ (for $i$ = 1, 2, ..., $N$; $j$ = 1, 2, ..., $M$)—Performance metric about alternatives.

Information given in matrix of decisions and a decision-making process is the decision-maker's duty to determine the best alternative and/or to make a ranking of alternatives. Normalization step is to be added to reach out at the same units [16] (Table 4.1).

### 4.3.3 Simple Additive Weighting (SAW) Method

SAW is also known as weighted sum method and it is normal and extensive aid of MADM method. Every attribute is assigned a weight and the summation of all weights must be 1. Regard to every attribute each alternative was estimated. Performance score of all composite alternative was estimated using Eq. 4.1

$$P_i = \sum_{j=1}^{M} w_j m_{ij} \tag{4.1}$$

After normalization of the decision table element, SAW capable of helping with any form and different attributes. For this, Eq. 4.2 will be used:

$$P_i = \sum_{j=1}^{M} w_j (m_{ij})_{\text{normal}} \tag{4.2}$$

**Table 4.1** Decision table in MADM methods

| Alternatives | Attributes | | | | | |
|---|---|---|---|---|---|---|
| | $B_1(w_1)$ | $B_2(w_2)$ | $B_3(w_3)$ | –(–) | –(–) | $B_M$ |
| $A_1$ | $m_{11}$ | $m_{12}$ | $m_{13}$ | – | – | $m_{1M}$ |
| $A_2$ | $m_{21}$ | $m_{22}$ | $m_{23}$ | – | – | $m_{2M}$ |
| – | – | – | – | – | – | – |
| – | – | – | – | – | – | – |
| $A_N$ | $m_{N1}$ | $m_{N2}$ | $m_{N3}$ | – | – | $m_{NM}$ |

where

$(m_{ij})_{normal}$ = normalized value of $m_{ij}$
$P_i$ = composite score of the alternative $A_i$

If there were no completion of summation of weights is equal to One, then Eq. 4.3 able to apply and this method knows as a simple multiple attribute rating technique (SMART) [16].

$$P_i = \left[ \sum_{j=1}^{M} w_j (m_{ij})_{normal} \right] / \sum_{j=1}^{M} w_j \qquad (4.3)$$

### 4.3.4   Weighted Product Method (WPM)

WPM is same as that of SAW. The main change is that multiplication was used in place of addition in the model. Composite Performance score of an alternative was calculated using Eq. 4.4

$$P_i = \prod_{j=1}^{M} \left[ (m_{ij})_{normal} \right]^{w_j} \qquad (4.4)$$

The method to normalize the values is same like SAW method. Each normalized value of an alternative with respect to an attribute, i.e. $(m_{ij})$ normal, is raised to the power of the relative weight of a respective attribute. Highest $P_i$ value shows that the considered alternative was the best alternative [17].

### 4.3.5   Analytical Hierarchy Process (AHP) Method

AHP, a popular analytical methodology applied to complex problems in decision making. This system is having capability to handle decision situations containing subjective judgments, multiple decision makers and have a capacity to monitor target continuity. Built to concentrate people's way of thinking, this approach appears to be highly regarded and the method of decision making applied more.

### 4.3.6  Procedure of Geometric Mean Method (AHP) is Given Below

**Step 1**: To estimate the objective as well as evaluation attributes. Form a hierarchy (objective, attributes and alternatives is at the top, second and third level respectively).

**Step 2**: Relative importance is to be determined with regard to the goal by considering parameters as different attributes.

From a matrix of pairwise contrast, considering scale of relative significance. Verdict is entered for analytic hierarchy process by using the fundamental scale. The value 1 is always assigned to an attribute when compared with itself, so all diagonal entries in a comparison matrix are 1. The numbers 3, 5, 7 and 9 support the moderate importance of the verbal evaluations, 'strong importance', 'very strong importance' and 'absolute importance' (with 2, 4, 6 and 8 for a concession between these values).

Where

$a_{ij}$ = Attribute $i$ is of comparative value with respect to attribute $j$
$M$ = Order of square matrix $B$

In below matrix, $b_{ij} = 1$ when $i = j$ and $b_{ji} = 1/b_{ij}$

$$B_{M \times M} = \begin{pmatrix} \text{Attributes} & B_1 & B_2 & B_3 & - & - & B_M \\ B_1 & 1 & b_{12} & b_{13} & - & - & b_{1M} \\ B_2 & b_{21} & 1 & b_{23} & - & - & b_{2M} \\ B_3 & b_{31} & b_{32} & 1 & - & - & b_{3M} \\ - & - & - & - & - & - & - \\ - & - & - & - & - & - & - \\ B_M & b_{M1} & b_{M2} & b_{M3} & - & - & 1 \end{pmatrix} \quad (4.5)$$

Calculate a normalized relative weight ($w_j$) for all attributes using

(i)  Finding the geometric mean respective of the $i$th row and
(ii) Normalization of geometric means of row values in the comparison matrix.

$$GM_j = \left[ \prod_{j=1}^{M} b_{ij} \right]^{1/M} \quad (4.6)$$

and

$$w_j = GM_j / \sum_{j=1}^{M} GM_j \quad (4.7)$$

Due to its simplicity, simple estimation of the maximum Eigen value and reduced variance of judgments, AHP is often used to calculate the relative normalized weights of the attributes.

Determine the $A3$ and $A4$ matrices as set out below:

$$A3 = A1 * A2$$
$$A4 = A3/A2,$$

where $A1 = B_{M \times M}$, $A2 = [w_1, w_2, w_3 \ldots w_j]^T$.

The average of matrix $A4$ gives the maximum value of Eigen $\lambda_{max}$.

Evaluate consistency index by using equation as $CI = (\lambda_{max} - M)/(M - 1)$. Values of CI results in a direct proportion to the variation from accuracy.

A random index (RI) is to be selected for the number of attributes used in decision making by using Table 4.2.

Consistency ratio $(CR) = CI/RI$ [CR $< = 0.1$ Mostly acceptable]

**Step 3**: Total or combined performance scores for an alternatives are calculated by taking product of respective standardized weight $(w_j)$ of each attribute (as per step 2) with its standardized weights for a particular values (calculated in step 2) and made summation of the attributes is same as that of SAW method [7] (Fig. 4.1).

Preference   Ranking   Organization   Method   for   Enrichment   Evaluations **(PROMETHEE)**

**Step 1**: For decision-making problem first, find the selection criteria and choose the alternatives based on the selected criteria which satisfies the need.

**Step 2**: After choosing the alternatives, from decision matrix have the measures or values for the selected alternatives of all the parameters. By the defined criteria, AHP method give weights of respective importance.

**Step 3**: Having an information based upon the decision-maker preference function, Users compare the contribution of the alternatives with respect to every individual criterion.

**Step 4**: Determine partial binary relationships showing the preference power of one alternative respectively of others by using pairwise comparison of alternatives in each criterion to.

Where, $P_i$ = Preference function, $w_i$ = weight for each criterion.

**Table 4.2** Random index (RI) values

| Attributes | 3 | 4 | 5 | 6 | 7 | 8 | 9 | 10 |
|---|---|---|---|---|---|---|---|---|
| RI | 0.52 | 0.89 | 1.11 | 1.25 | 1.35 | 1.4 | 1.45 | 1.49 |

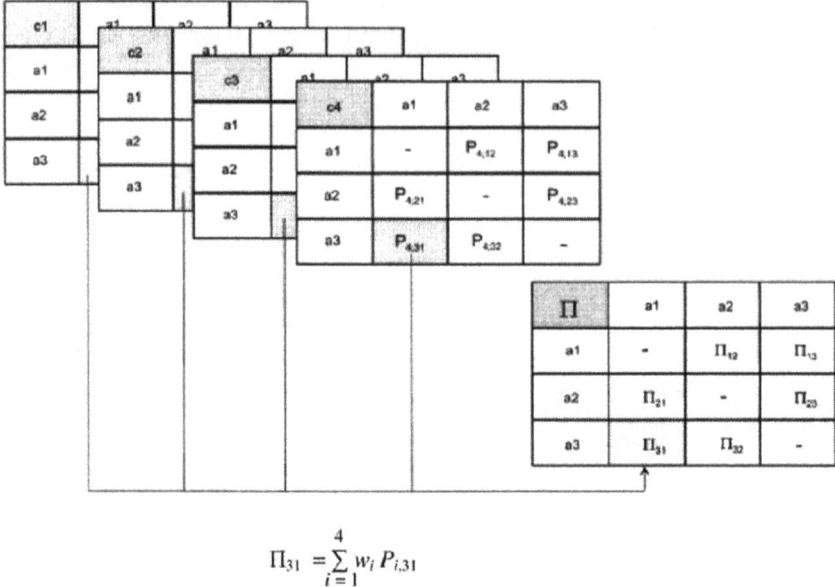

$$\Pi_{31} = \sum_{i=1}^{4} w_i\, P_{i,31}$$

**Fig. 4.1** Preference indexes for a problem with three alternatives and four parameters

**Step 5**: The final ranking is decided based on the descending order of Net Flow ($\varphi$).

$\varphi_+$ = sum of all elements of $i$th row $\varphi_-$ = sum of all elements of $i$th column

$$\varphi = \text{Net Flow} = \{\varphi_+\} - \{\varphi_-\}$$

It gives an overall ranking of the alternative in order of best to the worst with the aid of net flows [7].

### 4.3.7  Grey Relational Analysis (GRA)

In the GRA technique, at beginning data obtained from experiment/trial were normalized by taking into consideration of beneficial and non-beneficial attributes. Grey relational coefficients were determined from normalized experimental data [18]. The grey relational coefficient determines the association of actual and desired experimental data. Grey relational grades were computed by taking average of the grey relational coefficients [18] of each response variable (3 responses). Assessment of multi-process response variable depends upon grey relation grades [18] and is used to optimize process parameters.

**Step 1**: Data normalization is an important step because of different units and ranges of process parameters. For Beneficial attributes (higher the better), Eq. 4.8 is used to normalize the corresponding attributes.

$$X_i(k) = \frac{y_i(k) - \text{Min } y_i(k)}{\text{Max } y_i(k) - \text{Min } y_i(k)} \tag{4.8}$$

For Non-Beneficial attributes (lower the better), Eq. 4.9 is used to normalize the corresponding attributes.

$$X_i(k) = \frac{\text{Max } y_i(k) - y_i(k)}{\text{Max } y_i(k) - \text{Min } y_i(k)} \tag{4.9}$$

where

$x_i(k)$—value after the grey relational generation
Min $y_i(k)$—smallest value of $y_i(k)$ for the $k$th response
Max $y_i(k)$—largest value of $y_i(k)$ for the $k$th response

**Step 2**: Grey Relational Coefficient ($\xi_i$) is determined by using Eq. 4.10

$$\xi_i(k) = \frac{\nabla_{\min} + \varsigma * \nabla_{\max}}{\nabla O_i(k) + \varsigma * \nabla_{\max}} \tag{4.10}$$

where

$\nabla O_i = \|x_0(k) - x_i(k)\|$ = Deviation sequence = difference of absolute value between $x_0(k)$ and $x_i(k)$
$\varsigma$ = distinguishing coefficient (0–1)
$\Delta_{\min}$ = smallest value of $\Delta O_i$
$\Delta_{\max}$ = largest value of $\Delta O_i$.

**Step 3**: Grey relational grades ($Y_i$) computed by taking average of the grey relational coefficients shown in Eq. 4.11

$$'Y_i = \frac{1}{n} \sum_{k=1}^{n} w_k(k) * \xi_i(k) \tag{4.11}$$

where

$n$ = number of response variable
$w_k$ = weight related to each attribute

Higher value of grey relational grade results in an accurate degree of relation [18] (closer to optimal) between the $x_0(k)$ reference sequence and the $x_i(k)$ sequence given.

## 4.4   Case Study and Data Collection

### 4.4.1   Problem Definition

The problem of optimisation mentioned is on the basis of I-optimality criteria applied to FDM process parameters. This study also leads to formulation of mathematical equation to build a nonlinear relationship of process parameters and dimensional accuracy [17]. Parameters such as Layer thickness 'A' (mm), Air gap 'B' (mm), Raster angle 'C' (degree), Built orientation 'D' (degree), Road width 'E' (mm) and Number of counters 'F' are selected as input variables and percentage changes in length (L), width (W) and thickness (T) are selected as I-optimal design variables in each working environment category shown in Table 4.3.

Subject to constraints of input variables,

$$0.1270 \leq A \leq 0.3302$$

$$0 \leq B \leq 0.5$$

$$0 \leq C \leq 90$$

$$0 \leq D \leq 90$$

$$0.4572 \leq E \leq 0.5782$$

$$1 \leq F \leq 10$$

### 4.4.2   Analysis

Total of 60 cubic specimens were manufactured of size (35 mm × 12.5 mm 3.5 mm) using FDM Fortus 400 machine with test specimen of PC-ABS material used [17].

The optimized results shown by the earlier researcher are as follows:

The percentage change in length was decreased with changes in the raster angle from 0° to 90° and the percentage change in width of the component decreases linearly with decreases in layer thickness, air gap, road width and number of contours And it was sensibly enriched with an improvement in raster angle and build orientation from 0° to 90°. After application of SAW, WPM and PROMETHEE method to above-collected data the ranking achieved is as shown in Table 4.4.

**Table 4.3** Optimal design matrix and collected data

| Run | A | B | C | D | E | | $\Delta L$ | $\Delta W$ | $\Delta T$ |
|---|---|---|---|---|---|---|---|---|---|
| 1 | 0.254 | 0.5 | 90 | 90 | 0.4572 | 1 | 0.1714 | 0.528 | 2.1905 |
| 2 | 0.254 | 0.3 | 45 | 45 | 0.5298 | 5 | 0.1886 | 0.752 | 4.3429 |
| 3 | 0.254 | 0.3 | 45 | 45 | 0.5298 | 5 | 0.2 | 0.72 | 4.3686 |
| 4 | 0.3302 | 0.4 | 90 | 30 | 0.4814 | 1 | 0.1343 | 0.64 | 6 |
| 5 | 0.127 | 0.2 | 45 | 90 | 0.4572 | 1 | 0.0714 | 0.12 | 2.5714 |
| 6 | 0.254 | 0.2 | 45 | 90 | 0.4572 | 5 | 0.1571 | 0.52 | 5.0286 |
| 7 | 0.127 | 0 | 0 | 45 | 0.4572 | 5 | 0.0786 | 0.1064 | 4.2857 |
| 8 | 0.3302 | 0.5 | 0 | 90 | 0.4814 | 1 | 0.0571 | 0.62 | 8.2857 |
| 9 | 0.3302 | 0 | 90 | 90 | 0.4572 | 8 | 0.0571 | 0.304 | 12.38 |
| 10 | 0.127 | 0 | 45 | 0 | 0.5056 | 7 | 0.12 | 0.34 | 4.8571 |
| 11 | 0.3302 | 0.5 | 90 | 0 | 0.5782 | 1 | 0.18 | 0.66 | 6.0571 |
| 12 | 0.3302 | 0.2 | 0 | 90 | 0.4572 | 10 | 0.1029 | 0.54 | 10.1143 |
| 13 | 0.3302 | 0.3 | 90 | 0 | 0.4572 | 7 | 0.1 | 0.688 | 7.2857 |
| 14 | 0.1778 | 0.5 | 0 | 0 | 0.4572 | 1 | 0.0571 | 0.4 | 3.8286 |
| 15 | 0.127 | 0 | 90 | 45 | 0.4572 | 10 | 0.0143 | 0.14 | 4.6857 |
| 16 | 0.127 | 0.5 | 90 | 75 | 0.5056 | 5 | 0.1 | 0.16 | 2.5714 |
| 17 | 0.127 | 0.5 | 90 | 0 | 0.5782 | 10 | 0.0794 | 0.34 | 2.6286 |
| 18 | 0.3302 | 0 | 90 | 60 | 0.4814 | 1 | 0.0857 | 0.48 | 11.18 |
| 19 | 0.127 | 0.2 | 45 | 45 | 0.5782 | 10 | 0.1071 | 0.3 | 2.3429 |
| 20 | 0.127 | 0.5 | 30 | 75 | 0.554 | 8 | 0.0571 | 0.12 | 3.9143 |
| 21 | 0.3302 | 0 | 30 | 0 | 0.4572 | 1 | 0.1357 | 0.28 | 7.7143 |
| 22 | 0.254 | 0.5 | 90 | 90 | 0.4572 | 10 | 0.1429 | 0.576 | 3.2886 |
| 23 | 0.127 | 0 | 90 | 90 | 0.5782 | 3 | 0.0943 | 0.24 | 3.4286 |
| 24 | 0.3302 | 0 | 30 | 45 | 0.5298 | 10 | 0.0657 | 0.6 | 12.2857 |
| 25 | 0.1778 | 0 | 90 | 0 | 0.5298 | 1 | 0.1714 | 0.3 | 4 |
| 26 | 0.254 | 0.5 | 90 | 75 | 0.554 | 1 | 0.2071 | 0.66 | 2.9714 |
| 27 | 0.1778 | 0.5 | 60 | 0 | 0.4572 | 10 | 0.0571 | 0.56 | 3.2571 |
| 28 | 0.1778 | 0 | 0 | 90 | 0.5056 | 1 | 0.1514 | 0.22 | 5.2571 |
| 29 | 0.254 | 0.3 | 45 | 45 | 0.5298 | 5 | 0.1714 | 0.72 | 4.8571 |
| 30 | 0.127 | 0.3 | 0 | 0 | 0.5298 | 7 | 0.0857 | 0.24 | 3.4286 |
| 31 | 0.3302 | 0.5 | 45 | 45 | 0.4572 | 10 | 0.1429 | 0.8 | 8.9714 |
| 32 | 0.3302 | 0.5 | 90 | 90 | 0.5782 | 10 | 0.0286 | 0.6 | 8.6286 |
| 33 | 0.3302 | 0.5 | 0 | 0 | 0.5782 | 10 | 0.0571 | 0.768 | 9.7143 |
| 34 | 0.3302 | 0.2 | 60 | 90 | 0.5782 | 1 | 0.1314 | 0.664 | 8.5143 |
| 35 | 0.254 | 0.5 | 0 | 90 | 0.5782 | 7 | 0.0571 | 0.608 | 4.5714 |
| 36 | 0.3302 | 0.2 | 0 | 0 | 0.5782 | 1 | 0.0571 | 0.52 | 7.0286 |
| 37 | 0.3302 | 0 | 0 | 60 | 0.5782 | 5 | 0.1071 | 0.712 | 11.4286 |
| 38 | 0.254 | 0.3 | 45 | 45 | 0.5298 | 5 | 0.1857 | 0.736 | 4.8 |
| 39 | 0.127 | 0.5 | 0 | 90 | 0.5782 | 1 | 0.0929 | 0.16 | 3.4857 |
| 40 | 0.254 | 0.3 | 45 | 45 | 0.5298 | 5 | 0.1857 | 0.752 | 4.3429 |

(continued)

**Table 4.3** (continued)

| Run | A | B | C | D | E | | ΔL | ΔW | ΔT |
|-----|---|---|---|---|---|---|------|------|------|
| 41 | 0.254 | 0 | 45 | 90 | 0.5056 | 7 | 0.1143 | 0.48 | 8.5 |
| 42 | 0.127 | 0 | 45 | 0 | 0.5056 | 7 | 0.0857 | 0.18 | 4.1143 |
| 43 | 0.1778 | 0.5 | 45 | 30 | 0.5782 | 1 | 0.0929 | 0.48 | 2.9143 |
| 44 | 0.127 | 0.2 | 45 | 45 | 0.5782 | 10 | 0.0857 | 0.16 | 3.4057 |
| 45 | 0.127 | 0 | 0 | 0 | 0.5782 | 1 | 0.0571 | 0.24 | 2.7143 |
| 46 | 0.254 | 0 | 0 | 0 | 0.4572 | 10 | 0.0657 | 0.672 | 5.2571 |
| 47 | 0.127 | 0.3 | 90 | 0 | 0.4572 | 1 | 0.2286 | 0.08 | 3.1429 |
| 48 | 0.254 | 0.3 | 90 | 30 | 0.5056 | 10 | 0.1643 | 0.72 | 4.8571 |
| 49 | 0.1778 | 0 | 0 | 90 | 0.5782 | 10 | 0.1143 | 0.32 | 4.6571 |
| 50 | 0.1778 | 0.3 | 90 | 90 | 0.5298 | 10 | 0.0857 | 0.2664 | 2.5714 |
| 51 | 0.254 | 0.1 | 90 | 30 | 0.5782 | 5 | 0.1857 | 0.784 | 4.8 |
| 52 | 0.254 | 0.3 | 45 | 45 | 0.5298 | 5 | 0.1786 | 0.704 | 4.4571 |
| 53 | 0.3302 | 0.2 | 60 | 90 | 0.5782 | 1 | 0.072 | 0.616 | 8.8571 |
| 54 | 0.3302 | 0 | 90 | 0 | 0.5782 | 10 | 0.0857 | 0.816 | 11.6 |
| 55 | 0.254 | 0.3 | 0 | 45 | 0.4572 | 5 | 0.1914 | 0.72 | 4.5714 |
| 56 | 0.127 | 0.3 | 0 | 45 | 0.5056 | 10 | 0.0714 | 0.28 | 3.6514 |
| 57 | 0.3302 | 0.5 | 45 | 0 | 0.5056 | 5 | 0.0857 | 0.76 | 7.1429 |
| 58 | 0.127 | 0.4 | 15 | 60 | 0.5056 | 3 | 0.0762 | 0.16 | 4.1143 |
| 59 | 0.127 | 0.5 | 15 | 90 | 0.4572 | 10 | 0.0571 | 0.096 | 3.1057 |
| 60 | 0.254 | 0.3 | 45 | 45 | 0.5298 | 5 | 0.2 | 0.72 | 4.8571 |

Best result given by the earlier researcher in this work is shown in below Table 4.5, which are taken as reference (best solution) value for the calculation of combined percentage error of each response variable of sorted run and its rank shown in Table 4.9.

## 4.4.3 Application of Grey Relational Analysis (GRA)

All the response variables selected for the optimizations in above experiment are non-beneficial. The normalization of three response variables are done by applying Eq. 4.9 to Table 4.3 is shown in Table 4.6.

Grey Relational Coefficient ($\xi_i$) can be calculated for normalized values in Table 4.6 by using Eq. 4.10 is shown in Table 4.7.

$\nabla o_i = ||x_o(k) - x_i(k)||$ = Deviation sequence, $x_0(k)$ is taken as 1 (maximum values in the table), while $x_i(k)$ is taken as each normalized value corresponding to particular run ($i = 1, 2, 3, \ldots, 59, 60$). A value of $\varsigma$ (distinguishing coefficient) is small and distinguished ability is higher $\varsigma = 0.5$ is mostly used. The value of $\Delta_{min}$ is taken as 0 (zero) and $\Delta_{max}$ as 1.

**Table 4.4** Ranking of design matrix

| Run | A | B | C | D | E | F | SAW | WPM | PROMETHEE |
|-----|------|-----|----|----|--------|----|-----|-----|-----------|
| 1 | 0.254 | 0.5 | 90 | 90 | 0.4572 | 1 | 14 | 27 | 23 |
| 2 | 0.254 | 0.3 | 45 | 45 | 0.5298 | 5 | 35 | 50 | 52 |
| 3 | 0.254 | 0.3 | 45 | 45 | 0.5298 | 5 | 37 | 51 | 50 |
| 4 | 0.3302 | 0.4 | 90 | 30 | 0.4814 | 1 | 47 | 44 | 43 |
| 5 | 0.127 | 0.2 | 45 | 90 | 0.4572 | 1 | 4 | 3 | 2 |
| 6 | 0.254 | 0.2 | 45 | 90 | 0.4572 | 5 | 38 | 37 | 40 |
| 7 | 0.127 | 0 | 0 | 45 | 0.4572 | 5 | 7 | 5 | 8 |
| 8 | 0.3302 | 0.5 | 0 | 90 | 0.4814 | 1 | 43 | 34 | 30 |
| 9 | 0.3302 | 0 | 90 | 90 | 0.4572 | 8 | 33 | 30 | 25 |
| 10 | 0.127 | 0 | 45 | 0 | 0.5056 | 7 | 30 | 26 | 32 |
| 11 | 0.3302 | 0.5 | 90 | 0 | 0.5782 | 1 | 51 | 56 | 53 |
| 12 | 0.3302 | 0.2 | 0 | 90 | 0.4572 | 10 | 56 | 46 | 42 |
| 13 | 0.3302 | 0.3 | 90 | 0 | 0.4572 | 7 | 52 | 43 | 44 |
| 14 | 0.1778 | 0.5 | 0 | 0 | 0.4572 | 1 | 22 | 19 | 13 |
| 15 | 0.127 | 0 | 90 | 45 | 0.4572 | 10 | 1 | 1 | 5 |
| 16 | 0.127 | 0.5 | 90 | 75 | 0.5056 | 5 | 5 | 7 | 7 |
| 17 | 0.127 | 0.5 | 90 | 0 | 0.5782 | 10 | 13 | 15 | 11 |
| 18 | 0.3302 | 0 | 90 | 60 | 0.4814 | 1 | 54 | 39 | 36 |
| 19 | 0.127 | 0.2 | 45 | 45 | 0.5782 | 10 | 9 | 17 | 17 |
| 20 | 0.127 | 0.5 | 30 | 75 | 0.554 | 8 | 6 | 4 | 4 |
| 21 | 0.3302 | 0 | 30 | 0 | 0.4572 | 1 | 39 | 33 | 34 |
| 22 | 0.254 | 0.5 | 90 | 90 | 0.4572 | 10 | 24 | 32 | 29 |
| 23 | 0.127 | 0 | 90 | 90 | 0.5782 | 3 | 18 | 18 | 19 |
| 24 | 0.3302 | 0 | 30 | 45 | 0.5298 | 10 | 55 | 42 | 37 |
| 25 | 0.1778 | 0 | 90 | 0 | 0.5298 | 1 | 25 | 28 | 26 |
| 26 | 0.254 | 0.5 | 90 | 75 | 0.554 | 1 | 23 | 36 | 38 |
| 27 | 0.1778 | 0.5 | 60 | 0 | 0.4572 | 10 | 21 | 20 | 14 |
| 28 | 0.1778 | 0 | 0 | 90 | 0.5056 | 1 | 27 | 25 | 31 |
| 29 | 0.254 | 0.3 | 45 | 45 | 0.5298 | 5 | 42 | 48 | 50 |
| 30 | 0.127 | 0.3 | 0 | 0 | 0.5298 | 7 | 17 | 14 | 15 |
| 31 | 0.3302 | 0.5 | 45 | 45 | 0.4572 | 10 | 59 | 60 | 60 |
| 32 | 0.3302 | 0.5 | 90 | 90 | 0.5782 | 10 | 26 | 22 | 27 |
| 33 | 0.3302 | 0.5 | 0 | 0 | 0.5782 | 10 | 48 | 38 | 41 |
| 34 | 0.3302 | 0.2 | 60 | 90 | 0.5782 | 1 | 57 | 57 | 47 |
| 35 | 0.254 | 0.5 | 0 | 90 | 0.5782 | 7 | 28 | 23 | 22 |
| 36 | 0.3302 | 0.2 | 0 | 0 | 0.5782 | 1 | 32 | 29 | 24 |
| 37 | 0.3302 | 0 | 0 | 60 | 0.5782 | 5 | 60 | 59 | 55 |
| 38 | 0.254 | 0.3 | 45 | 45 | 0.5298 | 5 | 44 | 53 | 56 |
| 39 | 0.127 | 0.5 | 0 | 90 | 0.5782 | 1 | 12 | 10 | 16 |
| 40 | 0.254 | 0.3 | 45 | 45 | 0.5298 | 5 | 34 | 49 | 49 |

(continued)

**Table 4.4** (continued)

| Run | A | B | C | D | E | F | SAW | WPM | PROMETHEE |
|-----|------|-----|----|----|--------|----|-----|-----|-----------|
| 41 | 0.254 | 0 | 45 | 90 | 0.5056 | 7 | 53 | 41 | 39 |
| 42 | 0.127 | 0 | 45 | 0 | 0.5056 | 7 | 16 | 13 | 18 |
| 43 | 0.1778 | 0.5 | 45 | 30 | 0.5782 | 1 | 20 | 21 | 20 |
| 44 | 0.127 | 0.2 | 45 | 45 | 0.5782 | 10 | 11 | 8 | 9 |
| 45 | 0.127 | 0 | 0 | 0 | 0.5782 | 1 | 8 | 6 | 3 |
| 46 | 0.254 | 0 | 0 | 0 | 0.4572 | 10 | 31 | 31 | 33 |
| 47 | 0.127 | 0.3 | 90 | 0 | 0.4572 | 1 | 3 | 11 | 21 |
| 48 | 0.254 | 0.3 | 90 | 30 | 0.5056 | 10 | 41 | 47 | 48 |
| 49 | 0.1778 | 0 | 0 | 90 | 0.5782 | 10 | 29 | 24 | 28 |
| 50 | 0.1778 | 0.3 | 90 | 90 | 0.5298 | 10 | 10 | 12 | 6 |
| 51 | 0.254 | 0.1 | 90 | 30 | 0.5782 | 5 | 45 | 54 | 59 |
| 52 | 0.254 | 0.3 | 45 | 45 | 0.5298 | 5 | 36 | 45 | 45 |
| 53 | 0.3302 | 0.2 | 60 | 90 | 0.5782 | 1 | 50 | 35 | 35 |
| 54 | 0.3302 | 0 | 90 | 0 | 0.5782 | 10 | 58 | 58 | 57 |
| 55 | 0.254 | 0.3 | 0 | 45 | 0.4572 | 5 | 40 | 52 | 54 |
| 56 | 0.127 | 0.3 | 0 | 45 | 0.5056 | 10 | 19 | 16 | 12 |
| 57 | 0.3302 | 0.5 | 45 | 0 | 0.5056 | 5 | 49 | 40 | 46 |
| 58 | 0.127 | 0.4 | 15 | 60 | 0.5056 | 3 | 15 | 9 | 10 |
| 59 | 0.127 | 0.5 | 15 | 90 | 0.4572 | 10 | 2 | 2 | 1 |
| 60 | 0.254 | 0.3 | 45 | 45 | 0.5298 | 5 | 46 | 55 | 58 |

**Table 4.5** Best parameter setting and responses

| Run | Layer thickness (A) | Air gap (B) | Raster angle (C) | Built orientation (D) | Road width (E) | No. of contour (F) | $\Delta L$ | $\Delta W$ | $\Delta T$ |
|-----|------|------|--------|--------|-------|-------|------|------|------|
| Best solution | 0.127 | 0.342 | 88.918 | 89.122 | 0.462 | 1 | 0.056 | 0.027 | 1.978 |

Finally the grey relational grades $Y_i$ are calculated making average of the grey relational coefficients with Eq. 4.11 is as shown in Table 4.8. The value of rank is also calculated by putting criteria as higher the value of grey relational grade leads to higher rank.

The value of process responses ($n$) is taken as 3 and weight ($w_k$) related to each attribute is taken as 1 (similar weight). The difference in value of grey relational grade ($Y_i$) for rank 1 and 2 is of 0.004 (approximately equal to zero). Due to this small difference (0.004) the run 15 and run 59 is showing different ranks. But the grey relational grade value is approximately equal for both runs (15 and 59), hence the value of both runs are closer to optimal values.

**Table 4.6** Normalized data by using GRA

| Run | ΔL | ΔW | ΔT | Run | ΔL | ΔW | ΔT |
|-----|------|------|------|-----|------|------|------|
| 1 | 0.2669 | 0.3913 | 1 | 31 | 0.3999 | 0.0217 | 0.3345 |
| 2 | 0.1866 | 0.0869 | 0.7887 | 32 | 0.9332 | 0.2934 | 0.3681 |
| 3 | 0.1334 | 0.1304 | 0.7862 | 33 | 0.8002 | 0.0652 | 0.2616 |
| 4 | 0.4400 | 0.2391 | 0.6261 | 34 | 0.4535 | 0.2065 | 0.3793 |
| 5 | 0.7335 | 0.9456 | 0.9626 | 35 | 0.8002 | 0.2826 | 0.7663 |
| 6 | 0.3336 | 0.4021 | 0.7214 | 36 | 0.8002 | 0.4021 | 0.5251 |
| 7 | 0.6999 | 0.9641 | 0.7943 | 37 | 0.5669 | 0.1413 | 0.0933 |
| 8 | 0.8002 | 0.2663 | 0.4018 | 38 | 0.2001 | 0.1086 | 0.7439 |
| 9 | 0.8002 | 0.6956 | 0 | 39 | 0.6332 | 0.8913 | 0.8728 |
| 10 | 0.5067 | 0.6467 | 0.7382 | 40 | 0.2001 | 0.0869 | 0.7887 |
| 11 | 0.2267 | 0.2119 | 0.6205 | 41 | 0.5333 | 0.4565 | 0.3807 |
| 12 | 0.5865 | 0.375 | 0.2223 | 42 | 0.6668 | 0.86413 | 0.8111 |
| 13 | 0.6000 | 0.1739 | 0.4999 | 43 | 0.6332 | 0.4565 | 0.9289 |
| 14 | 0.8002 | 0.5652 | 0.8392 | 44 | 0.6668 | 0.8913 | 0.8807 |
| 15 | 1 | 0.9184 | 0.7551 | 45 | 0.8002 | 0.7826 | 0.9485 |
| 16 | 0.6000 | 0.8913 | 0.9626 | 46 | 0.7601 | 0.1956 | 0.6990 |
| 17 | 0.69622 | 0.6467 | 0.9570 | 47 | 0 | 1 | 0.9065 |
| 18 | 0.6668 | 0.4565 | 0.1177 | 48 | 0.3000 | 0.1304 | 0.7382 |
| 19 | 0.5669 | 0.7010 | 0.9850 | 49 | 0.5333 | 0.6739 | 0.7579 |
| 20 | 0.80028 | 0.9456 | 0.8308 | 50 | 0.6668 | 0.7467 | 0.9626 |
| 21 | 0.4335 | 0.7282 | 0.4578 | 51 | 0.2001 | 0.0434 | 0.7439 |
| 22 | 0.3999 | 0.3260 | 0.8922 | 52 | 0.2333 | 0.1521 | 0.7775 |
| 23 | 0.6266 | 0.7826 | 0.878 | 53 | 0.7307 | 0.2717 | 0.3457 |
| 24 | 0.7601 | 0.2934 | 0.0092 | 54 | 0.6668 | 0 | 0.0765 |
| 25 | 0.2669 | 0.7010 | 0.822 | 55 | 0.1735 | 0.1304 | 0.7663 |
| 26 | 0.1003 | 0.2119 | 0.9233 | 56 | 0.7335 | 0.7282 | 0.8566 |
| 27 | 0.8002 | 0.3478 | 0.8953 | 57 | 0.6668 | 0.0760 | 0.5139 |
| 28 | 0.3602 | 0.8097 | 0.6990 | 58 | 0.7111 | 0.8913 | 0.8111 |
| 29 | 0.2669 | 0.1304 | 0.7382 | 59 | 0.8002 | 0.9782 | 0.9101 |
| 30 | 0.6668 | 0.7826 | 0.8780 | 60 | 0.1334 | 0.1304 | 0.7382 |

## 4.4.4   Results and Discussion

By comparing above ranking the sorting (rank 1–10) and combined average percentage error is determined as shown in Table 4.9. This table also gives the ranking of sorted run on the basis of combined average percentage error. This percentage error is calculated for each response variable and at the end, average of all the percentage error is calculated. Then best rank is calculated for lower combined average percentage error.

The results comparing the above-mentioned methods are found to be.

**Table 4.7** Grey relational coefficient ($\xi_i$)

| Run | $\xi_i$ ($\Delta L$) | $\xi_i$ ($\Delta W$) | $\xi_i$ ($\Delta T$) | Run | $\xi i$ ($\Delta L$) | $\xi_i$ ($\Delta W$) | $\xi_i$ ($\Delta T$) |
|-----|------|------|------|-----|------|------|------|
| 1 | 0.4054 | 0.4509 | 1 | 31 | 0.4545 | 0.3382 | 0.4290 |
| 2 | 0.3807 | 0.3538 | 0.703 | 32 | 0.8822 | 0.4144 | 0.4417 |
| 3 | 0.3658 | 0.3650 | 0.7005 | 33 | 0.7145 | 0.3484 | 0.4037 |
| 4 | 0.4717 | 0.3965 | 0.572 | 34 | 0.4778 | 0.3865 | 0.4461 |
| 5 | 0.6523 | 0.9019 | 0.930 | 35 | 0.7145 | 0.4107 | 0.681 |
| 6 | 0.4286 | 0.4554 | 0.642 | 36 | 0.7145 | 0.4554 | 0.5129 |
| 7 | 0.6249 | 0.9330 | 0.7085 | 37 | 0.535 | 0.368 | 0.3554 |
| 8 | 0.7145 | 0.4052 | 0.455 | 38 | 0.3846 | 0.3593 | 0.6612 |
| 9 | 0.7145 | 0.6216 | 0.333 | 39 | 0.5768 | 0.8214 | 0.7973 |
| 10 | 0.5034 | 0.5859 | 0.656 | 40 | 0.3846 | 0.3538 | 0.703 |
| 11 | 0.3927 | 0.3881 | 0.5685 | 41 | 0.5172 | 0.4791 | 0.4467 |
| 12 | 0.5473 | 0.4444 | 0.391 | 42 | 0.6001 | 0.7863 | 0.7258 |
| 13 | 0.5556 | 0.3770 | 0.4999 | 43 | 0.5768 | 0.4791 | 0.8756 |
| 14 | 0.7145 | 0.5348 | 0.7567 | 44 | 0.6001 | 0.8214 | 0.8074 |
| 15 | 1 | 0.8598 | 0.6712 | 45 | 0.7145 | 0.69697 | 0.9067 |
| 16 | 0.5556 | 0.8214 | 0.9304 | 46 | 0.6758 | 0.3833 | 0.6242 |
| 17 | 0.6220 | 0.5859 | 0.9208 | 47 | 0.3333 | 1 | 0.8425 |
| 18 | 0.6001 | 0.4791 | 0.3617 | 48 | 0.4166 | 0.3650 | 0.6564 |
| 19 | 0.5358 | 0.6258 | 0.9709 | 49 | 0.5172 | 0.6052 | 0.6737 |
| 20 | 0.7145 | 0.9019 | 0.7471 | 50 | 0.6001 | 0.6637 | 0.9304 |
| 21 | 0.4688 | 0.6478 | 0.4797 | 51 | 0.3846 | 0.3432 | 0.6612 |
| 22 | 0.4545 | 0.4259 | 0.8226 | 52 | 0.3947 | 0.3709 | 0.6920 |
| 23 | 0.5725 | 0.6969 | 0.8044 | 53 | 0.6499 | 0.4070 | 0.4331 |
| 24 | 0.6758 | 0.4144 | 0.3354 | 54 | 0.6001 | 0.3333 | 0.3512 |
| 25 | 0.4054 | 0.6258 | 0.7379 | 55 | 0.3769 | 0.3650 | 0.6815 |
| 26 | 0.3572 | 0.3881 | 0.8670 | 56 | 0.6523 | 0.6478 | 0.7771 |
| 27 | 0.7145 | 0.4339 | 0.8268 | 57 | 0.6001 | 0.3511 | 0.5070 |
| 28 | 0.4386 | 0.7244 | 0.6242 | 58 | 0.6338 | 0.8214 | 0.7258 |
| 29 | 0.4054 | 0.3650 | 0.6564 | 59 | 0.7145 | 0.9583 | 0.8477 |
| 30 | 0.6001 | 0.6969 | 0.8044 | 60 | 0.36588 | 0.3650 | 0.6564 |

PROMETHEE method gives first rank for Run 59, while SAW, WPM and GRA are showing second rank for the same run. As, these methods are showing different rank but the best rank given by combined average percentage error is also for run 59. This indicates that's the run 59 is confirmed best combination shown, As these ranks are calculated on the basis of percentage error in comparison with of best response (solution) shown in earlier study. And these results are validated by using GRA method as well.

**Table 4.8** Grey relational grade ($Y_i$)

| Run | Grey relational grade ($Y_i$) | Rank | Run | Grey relational grade ($Y_i$) | Rank |
|-----|-------------------------------|------|-----|-------------------------------|------|
| 1 | 0.619 | 22 | 31 | 0.407 | 60 |
| 2 | 0.479 | 46 | 32 | 0.579 | 28 |
| 3 | 0.477 | 48 | 33 | 0.489 | 38 |
| 4 | 0.480 | 44 | 34 | 0.437 | 57 |
| 5 | 0.828 | 3 | 35 | 0.602 | 23 |
| 6 | 0.509 | 36 | 36 | 0.561 | 31 |
| 7 | 0.756 | 7 | 37 | 0.420 | 59 |
| 8 | 0.525 | 35 | 38 | 0.468 | 52 |
| 9 | 0.557 | 32 | 39 | 0.732 | 9 |
| 10 | 0.582 | 27 | 40 | 0.481 | 42 |
| 11 | 0.450 | 56 | 41 | 0.481 | 41 |
| 12 | 0.461 | 55 | 42 | 0.704 | 15 |
| 13 | 0.478 | 47 | 43 | 0.644 | 21 |
| 14 | 0.669 | 19 | 44 | 0.743 | 8 |
| 15 | 0.844 | 1 | 45 | 0.773 | 5 |
| 16 | 0.769 | 6 | 46 | 0.561 | 30 |
| 17 | 0.710 | 14 | 47 | 0.725 | 12 |
| 18 | 0.480 | 43 | 48 | 0.479 | 45 |
| 19 | 0.711 | 13 | 49 | 0.599 | 24 |
| 20 | 0.788 | 4 | 50 | 0.731 | 10 |
| 21 | 0.532 | 34 | 51 | 0.463 | 53 |
| 22 | 0.568 | 29 | 52 | 0.486 | 40 |
| 23 | 0.691 | 18 | 53 | 0.497 | 37 |
| 24 | 0.475 | 50 | 54 | 0.428 | 58 |
| 25 | 0.590 | 26 | 55 | 0.475 | 51 |
| 26 | 0.538 | 33 | 56 | 0.692 | 17 |
| 27 | 0.658 | 20 | 57 | 0.486 | 39 |
| 28 | 0.596 | 25 | 58 | 0.727 | 11 |
| 29 | 0.476 | 49 | 59 | 0.840 | 2 |
| 30 | 0.701 | 16 | 60 | 0.462 | 54 |

The best parametric combination shown by PROMETHEE method and that are confirmed with the best response percentage error shown are: 0.127 mm as layer thickness; 0.5 mm as air gap; 150 as raster angle; 900 as build orientation; 0.4572 mm as road width and 10 as no. of contour, gives lower combined average percentage error of response variables.

**Table 4.9** Bestranking (solution) of sorted design matrix

| Run | SAW | WPM | PROME TEHEE | GRA | $\Delta L$ | $\Delta W$ | $\Delta T$ | Combined avg. percentage error | Rank |
|-----|-----|-----|-------------|-----|------|------|--------|------|------|
| 5 | 4 | 3 | 2 | 3 | 0.0714 | 0.12 | 2.5714 | 133.98 | 2 |
| 7 | 7 | 5 | 8 | 7 | 0.0786 | 0.1064 | 4.2857 | 150.37 | 4 |
| 15 | 1 | 1 | 5 | 1 | 0.0143 | 0.14 | 4.6857 | 209.96 | 8 |
| 16 | 5 | 7 | 7 | 6 | 0.1 | 0.16 | 2.5714 | 200.39 | 6 |
| 19 | 9 | 17 | 17 | 13 | 0.1071 | 0.3 | 2.3429 | 373.60 | 13 |
| 20 | 6 | 4 | 4 | 4 | 0.0571 | 0.12 | 3.9143 | 148.10 | 3 |
| 39 | 12 | 10 | 16 | 9 | 0.0929 | 0.16 | 3.4857 | 211.57 | 9 |
| 44 | 11 | 8 | 9 | 8 | 0.0857 | 0.16 | 3.4057 | 205.94 | 7 |
| 45 | 8 | 6 | 3 | 5 | 0.0571 | 0.24 | 2.7143 | 276.03 | 11 |
| 47 | 3 | 11 | 21 | 12 | 0.2286 | 0.08 | 3.1429 | 187.80 | 5 |
| 50 | 10 | 12 | 6 | 10 | 0.0857 | 0.2664 | 2.5714 | 323.23 | 12 |
| 58 | 15 | 9 | 10 | 11 | 0.0762 | 0.16 | 4.1143 | 212.22 | 10 |
| 59 | 2 | 2 | 1 | 2 | 0.0571 | 0.096 | 3.1057 | 104.84 | 1 |

## 4.5   Conclusion

Under the study, it has represented various techniques for an effective FDM process parametric optimization with the help of SAW, WPM, PROMETHEE and GRA method. This study presents a successful application of multi-attribute decision-making methods to experimental data of FDM process parameter optimization. Following are the conclusions based on this study.

- In MADM methods, SAW and WPM is best suited for small and quantitative data set, while PROMETHEE method more applicable for large, qualitative and quantitative data set. So, PROMETHEE method is more comfortable for handling more attributes and alternatives in different experiments or array values.
- As SAW, WPM and GRA shown the highest rank for run 15, but the percentage change in width and thickness are showing more for run 15 as compared to run 59.
- Similarly PROMETHEE method has given the highest rank for run 59, which gives less percentage change in width and thickness as compared to run 15.
- Best rank shown by GRA is approximately equal to the best rank shown by PROMETHEE method; hence GRA is also showing a result which is closer to optimal one.
- With the aid of soft-computing MADM method is easily applicable for optimization of dimensional accuracy.

# References

1. Rao, R.V., Rai, D.P.: Optimization of fused deposition modelling process using teaching learning-based optimization algorithm. Eng. Sci. Technol. Int. J. **19**, 587–603 (2015)
2. Noriega, A., Blanco, D., Alvarez, B.J., Garcia, A.: Dimensional accuracy improvement of FDM square cross-section parts using artificial neural networks and an optimization algorithm. Int. J. Adv. Manuf. Technol. 2301–2313 (2013)
3. Rong-Ji, W., Xin-Hua, L., Qing-Ding, W.: Optimizing process parameters for selective laser sintering based on neural network and genetic algorithm. Int. J. Adv. Manuf. Technol. 1035–1042 (2008)
4. Rayegani, F., Onwubolu, G.C.: Fused deposition modeling (FDM) process parameter prediction and optimization using group method for data handling (GMDH) and differential evolution (DE). Int. J. Adv. Manuf. Technol. 509–519 (2014)
5. Sandhu, K., Singh, G., Singh, S., Kumar, R., Prakash, C., Ramakrishna, S., Królczyk, G., Pruncu, C.I.: Surface characteristics of machined polystyrene with 3D printed thermoplastic tool. Materials **13**(12), 2729 (2020)
6. Singh, S., Singh, G., Sandhu, K., Prakash, C., Singh, R.: Investigating the optimum parametric setting for MRR of expandable polystyrene machined with 3D printed end mill tool. Mater. Today. Proc. (2020)
7. Venkata Rao, R., Patel, B.K.: Decision making in the manufacturing environment using an improved PROMETHEE method. Int. J. Prod. Res. 4665–4682 (2010)
8. Rao, R.V.: Jaya: a simple and new optimization algorithm for solving constrained and unconstrained optimization problems. Int. J. Ind. Eng. Computations **7**, 19–34 (2016)
9. Yim, S.J., Lee, D.Y.: Comparative study of search methods for the scheduling of flexible manufacturing systems. IFAC Control Ind. Syst. 1427–1432 (1997)
10. Adinarayana, M., Prasanthi, G., Krishnaiah, G.: Optimization for surface roughness, MRR, power consumption in turning of EN24 alloy steel using genetic algorithm. Int. J. Mech. Eng. Rob. Res. **3**, 21–26 (2014)
11. Ramezan, A.M., Navid, K., Fakhrabadi, M.M.S.: Optimization of milling parameters using artificial neural network and artificial immune system. J. Mech. Sci. Technol. 4097–4104 (2012)
12. Chen, D., Zou, F., Li, Z., Wang, J., Li, S.: An improved teaching–learning-based optimization algorithm for solving global optimization problem. Inf. Sci. 171–190 (2014)
13. Mahmood, K., Karaulovaa, T., Ottawa, T., Shevtshenkoa, E.: Performance analysis of a flexible manufacturing system (FMS). Procedia CIRP 424–429 (2017)
14. Chaudhry, I.A., Khanb, A.A.: A research survey: a review of flexible job shop scheduling techniques. Int. Trans. Oper. Res. 551–591 (2015)
15. Venkataramanaish, S.: Scheduling in a cellular manufacturing system: a heuristic approach. Int. J. Prod. Res. 429–449 (2007)
16. Venkata Rao, R.: Decision Making in the Manufacturing Environment (Using Graph Theory and Fuzzy Multiple Attribute Decision Making Methods), Springer Series in Advanced Manufacturing. Springer-Verlag London Limited 2007. ISSN 1860–5168
17. Mohamed, O.A., Masood, S.H., Bhowmik, J.L.: Optimization of fused deposition modelling process parameters for dimensional accuracy using I-optimality criterion. Measurement **81**, 174–196 (2016)
18. Hasani, H., Tabatabaei, S.A., Amiri, G.: Grey Relational analysis to determine the optimum process parameters for open-end spinning yarns. J. Eng. Fibers Fabr. **7**, 81–86 (2012)

# Chapter 5
# A Bibliometric Analysis on 3D Printed Concrete in Architecture

Nur Banu Gülle and Semra Arslan Selçuk

**Abstract** 3D printing technologies with its advantages such as speed, precision, on-site direct printing, non-stop production, using an adequate amount of material, and manufacturing complex shapes, are effective laboratories for researchers working in field of architectural design. With these technologies, almost any kind of plasticized material can be used and many forms and structures can be printed from scale models to the one-to-one scale end products. With the introduction of concrete, the most widely used material in the construction industry, to the 3D printers, revolutionary developments have occurred in building construction. Although the studies proceed with an experimental approach, authorities argue that this form of production will shift the paradigm of building construction. This article discusses what kinds of change/transformation caused by the technological changes/ developments are experienced in the Construction 4.0 process, as one of the most important items of the construction industry: "building with concrete". This discussion has been conducted through the articles written in English between 2000 and 2020, collected with a comprehensive literature review of ISI Web of Science Database. Keyword scanning is limited with concrete/3D printing/digital production and articles containing a combination of these terms in the title/summary/ keywords have been included. In this context, cooperation networks between leading countries, institutions, and actors have been presented while the latest developments related to the use of concrete in 3D printers have been revealed in this bibliometric analysis, and it has been possible to show these networks through clustering technics. It is thought that, by tracing the correlations between the keywords, interfaces of research on forms, materials, technology, and architectural applications that need more research will also be determined.

**Keywords** Digital manufacturing · 3D printing · Digital concrete · 3D printed concrete · Bibliometric analysis

N. B. Gülle · S. A. Selçuk (✉)
Faculty of Architecture, Department of Architecture, Gazi University, Ankara, Turkey
e-mail: semraselcuk@gazi.edu.tr

© The Author(s), under exclusive license to Springer Nature Switzerland AG 2022    77
K. Sandhu et al. (eds.), *Sustainability for 3D Printing*, Springer Tracts
in Additive Manufacturing, https://doi.org/10.1007/978-3-030-75235-4_5

## 5.1 Introduction

Construction industry plays a key role in the economies of countries. According to a report published by the World Economic Forum, the construction industry currently covers 6% of the world's gross domestic product (GDP)[1] [1] and is expected to reach a ratio of 14.7% in 2030 [2]. The construction industry is a strategically important sector for the European Economy, involving a wide range of stakeholders and companies that employ 18 million individuals [3]. The construction industry, which has one of the most important impacts on the environment, has the highest rate in resource and raw material consumption.

Industry sectors such as automotive, aviation, and aerospace have undergone radical process changes, adopting digital technologies to improve quality and efficiency. This digital transformation usually defined as Industry 4.0. [4, 5] connects system production technologies and smart production processes and radically transforms industry, production value chains and business models. This industrial transformation is driven by the shift from physical to digital through the use of sensors and controls, augmented reality systems, cognitive and high-performance computing, additive manufacturing, advanced materials, design and simulation systems, autonomous robots, and digital electronics, among other technologies.

Unlike other industries, the construction industry falls behind to adopt new technologies and has never undergone a recent major transformation [6]. The uniqueness of the construction industry poses a challenge for direct adaptation of technologies utilized in other industries. As it was during the last decades, non-digital technologies developed by the Second Industrial Revolution and the Third Industrial Revolution are still being used in a rather inefficient manner in construction project implementations with a focus on manpower. The usage of innovative systems is quite limited, especially in a sector known for its relatively reserved structure to innovations compared to other sectors. Moreover, these technologies are yet to be used in most countries [7]. The perceived image of the construction industry is characterized by low performance and quality image, predominantly low-tech and currently based on craft-based methods [8, 9].

The 4th industrial revolution, Industry 4.0, is the most recent motive in smart automation technology. In this New Age, the usage of modern manufacturing skills as part of integrating new information technologies plays a substantial role in economic competitiveness. Industry 4.0 offers cyber and physical systems to collaborate profitably by redefining the role of humanity with an aim of building smart systems. Basic concepts related to the virtual environment includes such as the Internet of Things (IoT), Big Data, Cloud Computing, etc. as the physical space comprises autonomous robots and additive manufacturing [10]. On the other hand, the physical part of smart systems is limited by the capacity of existing production systems. This makes additive manufacturing one of the vital components of

---

[1]Gross domestic product; within a certain time within the borders of a country, produced all ultimate goods and your services currency unit is the value in terms.

Industry 4.0. Due to the need for mass customization in Industry 4.0, unconventional production methods need to be improved. Therefore, additive manufacturing is becoming a key technology for producing customized products thanks to its ability in creating sophisticated objects with advanced properties (new materials, forms). Additive manufacturing is now used directly in various industries such as aviation, biomedical, and manufacturing thanks to increased product quality. Although there are still some doubts about its applicability in mass production, the usage of additive manufacturing in the industry is increasing due to new technological advances. As an evolving technology to create accurate and reinforced complex objects with increased production speed, it is thought as a way to change traditional manufacturing techniques in the near future.

Similar to the changes in other manufacturing sectors, the concept called Construction 4.0 [11] is defined as the automation and digitalization of processes by using modern technologies and new production techniques that comprise Industry 4.0 throughout the construction project life cycle. It is thought that this transformation will allow construction companies to increase efficiency, reduce project delays and cost overruns, manage complexity, and improve safety, resource efficiency, and quality [12, 13].

Digital production technologies (particularly 3D printing) have become an integral part of modern-day production and this technology has been commercialized to a large extent. Similarly, in the concrete industry, the potential of new technologies to implement and the economic, technical, and environmental benefits they will bring to this sector are considerably high [14]. While research and development activities in this field especially the rheological properties of "fresh concrete", on-site robotic solutions, reinforcement integration, test methods, etc. continue to develop, there is a still need for in-depth research on relevant subjects. Interdisciplinary partnerships are pivotal to the development of practical solutions. Collaborative works among architects, material researchers, civil engineers, robotics experts, etc. are required. In this context, this study aims primarily to examine the methods and techniques that have been included in the literature on printable concrete together with their examples. Moreover, the aim is to examine academic studies written about printable concrete[2] to reveal the intellectual structure in the field by analyzing it with bibliometric methods. This article is important in terms of showing the architectural and engineering researchers from a perspective that reveals the printable concrete features. The results of the research can be a guide for researches on printable concrete and 3D printers.

This study handles researches "printable concrete" in the field of architecture and engineering in a wider framework. Thanks to bibliometric techniques, the study presents ideas about the areas in which the articles are clustered, what kind of changes they have undergone over the years and the collaborations in the field [15].

---

[2]Bibliometry is an innovative method used to systematically search the literature. The biggest advantage of this approach is that it can analyze several scientific publications written on a subject and reveal the general features of research in the field through visualization [14].

The relevant articles in the WOS database were analyzed using keywords and bibliometric methods in the R program within this study. Thus, the conceptual map of the articles on "printable concrete" was presented and the most prolific authors in the field and the articles with the highest contribution were determined. Bibliometric research makes it possible to analyze a large number of studies in a short time thanks to newly developed programs [16, 17]. Research findings can help researchers to better understand current and future research trends and thus design more innovative and creative research by demonstrating the importance of this topic in the construction industry. Based on these points, to determine the position of printable concrete research in architectural and engineering literature, questions such as;

- What are the digital concrete studies and applications in the field of architecture and construction?
- What are the descriptive data (publication, author, reference numbers, and related indexes) of printable concrete research conducted in the field of architecture and construction?
- What is the social and intellectual structure (collaborations, social networks, trends) of printable concrete research conducted in architecture and engineering fields?

was examined to look at the field from a panoramic perspective by searching for answers.

## 5.2 3D Printed Concrete

Construction industry experienced innovative processes in design, material, and engineering in the twentieth century. The technological development of concrete has been a great convenience for modern architecture. Today, concrete is still one of the most preferred materials to construct all kinds of simple or complex buildings. More than 9 billion tons of cement was exported to make concrete in 2011 [18, 19]. This is due to the natural features of concrete. These are its global availability as raw materials, relative ease of processing, the ability to switch from a liquid state to a solid state where it can fill a mold and structural load-bearing capacity.

Technological and industrial developments have enabled architects, engineers and, builders to construct increasingly complex architectural structures from concrete. Computer-aided design and manufacturing (CAD/CAM) techniques have recently become modernized and have increased the possibilities to realize increasingly complex geometries [20]. There are not many alternatives to reinforced concrete building applications today. This is due to it is the ability to realize the tasks requiring large amounts of material for the necessary properties such as carrying a structural load and being durable, with affordable costs. However, while structural engineers can design financially efficient shapes through design tools, the realization has been limited primarily by the higher costs of custom molds. That is to say, processing has a restrictive effect on shape freedom. The prevalence of linear walls, columns, and slabs is an outcome related to the efficiency of easily constructed and reusable molds. Digital manufacturing processes fundamentally change this by enabling construction to be performed more effectively without using traditional molds. Most complex forms produced with digital design tools are mostly irrelevant with the default production modes used in concrete construction today. A huge gap has emerged between the capabilities of digital production technology in architectural design and the reality of the construction industry, where efficient solutions for the production of complex concrete structures are not available. When the digital production technologies and new insights from materials science are combined, the existing concrete techniques can be modified to improve mold efficiency and develop waste reduction approaches in the construction of complex concrete structures [21].

Because of the limited size of the objects that can be produced and costly equipment and long production times, additive manufacturing processes have a very limited application in building design and production. They are mostly used for the production of models with complex, curvilinear geometries in design. They are used to manufacture components in series, such as steel elements in lightweight lattice structures, by forming molds that are used in construction and later in investment casting. However, there are several recent experimental examples based on sprayed concrete to produce large-scale building components directly from digital data [22]. Two of the 3D concrete architectures, whose numbers are increasing day by day, are included in this study. The first is the "Office of the Future", which was designed by Killa Design company and constructed in 2016. This 3D printed offices are designed for the United Arab Emirates National Committee. The 3D-printed office is a fully functional building containing electricity, plumbing, telecommunications, and air conditioning systems. This 3D printed structure was produced in China and the parts were sent to Dubai after being printed. Within the project, it was reported that there was a 50–80% reduction in labor costs and a 30–60% reduction in construction waste [23] (Fig. 5.1).

Another example is produced by Win Sun firm's development of the world's largest 3D construction printer. This printer has 10 m width, 6.6 m height, and 150 m length capacity. A 5-story apartment and 1100 square meters' villas were produced using recycled concrete with 3D printers in Suzhou, China (Fig. 5.2). Another aspect that makes the buildings special and important is the usage of a "patented concrete material" consisting of recycled construction site wastes in their construction and work duration of a short period as 24 h [25, 26].

These two examples clearly illustrate that the use of 3D concrete in the field of architecture has revolutionary potentials.

**Fig. 5.1** General view of the building [24]

**Fig. 5.2** The first apartment and villa produced with 3D printers [27, 28]

## 5.2.1  Digital Production Methods with Concrete

In this part of the study, studies on how concrete can be shaped and printed digitally is discussed. As it is known, various additive manufacturing technologies for concrete are currently being developed worldwide. According to Buswell, there are more than 30 techniques and technologies that are digitally developing for the placement and production of cement-based materials [29]. It is not feasible to cover all the different techniques in just one study. But, a brief review of digital production methods with concrete can be epitomized as follows. In general, concrete and digital production methods can be classified on a process basis (Fig. 5.3), but they all illustrate common applications of concrete construction methods without the use of traditional molds.

### 5.2.1.1  Extrusion

It can be argued that the most popular method for "3D printing with concrete" is extrusion. Most researchers focus on multi-layer extrusion, where pre-mixed material is accumulated layer by layer to form an object. Extrusion is a process in which material is pushed through a nozzle and placed by the print head in a specific position as the print head moves in cavity (Fig. 5.4). An object is constructed layer

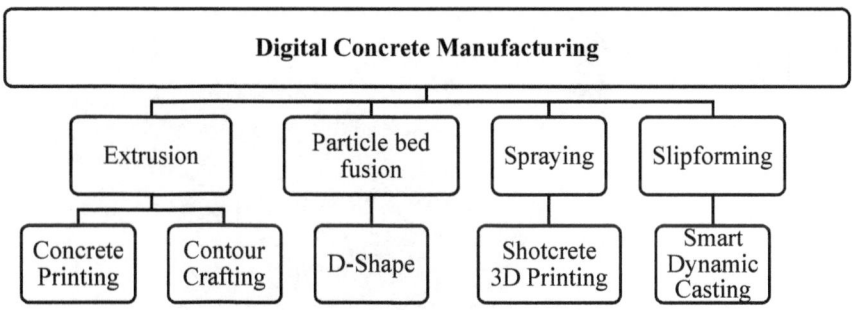

**Fig. 5.3** A Classification for large scale additive manufacturing processes with cement-based materials [30]

by layer with polymers similar to the well-known "fused deposition modeling" process. Portal-type printers in which a printhead is controlled in 3 axes on a fixed rectangular frame, robotic arms in which the print head is fixed to the end of a robot arm that typically has both translational and rotational and degrees of freedom and delta printers that have a hanging printhead with three retractable arms and thus additional freedom of rotation compared to portal-type printers [31, 32].

In the concrete extrusion process, moldable concrete is pressed continuously from a mold or nozzle of a predefined cross-section. The best-known multi-layer extrusion approaches are Concrete Printing and Contour Crafting.

- **Concrete Printing (3DCP)**

Concrete Printing (3BCP) is a layered extrusion method developed at Loughborough University [29, 34]. Unlike contour crafting, concrete printing is an off-site process aiming higher resolutions (layer thickness 4–6 mm) (Fig. 5.5). It allows better control of what is produced and greater geometric freedom. In 2018, the TU/e research group 3BCP [35], have built the world's first 3D printed, pre-stressed and, reinforced concrete bicycle bridge with their industrial partners [36]. Constructed in the Dutch village of Gemert, the bridge has a span of 6.5 m and a width of 3.5 m. It consists of six prefabricated segments with a height of approximately 1 m. The elements were rotated 90° after printing, glued and, pressed together with pre-tensioned tendons [30].

- **Contour Crafting**

Contour Crafting [37–39] is a layered extrusion technology developed at the University of Southern California by Behrokh Khoshnevis to automate the construction of structures of a complete house, including wiring, plumbing, drywall and, insulation (Fig. 5.6). Contour Crafting reduces construction costs, increases construction speed, provides more flexibility for architectural design and, offers a safe and sustainable environment. Construction cost can be estimated by the amount of material, time, and energy spent by the tool used to construct the building. When the total construction time is converted into a building model

**Fig. 5.4** A scheme of the additive extrusion process, the deposition of layers of a pre-mixed material. C, A, and W; cement, aggregate, and water respectively [33]

**Fig. 5.5** 3DCP technology [30]

stereolithography (STL) file, it can be followed generally after a tool path is determined [40].

A tool path of Contour Crafting expresses the position, direction, speed, and feed rate of the nozzle for the total construction time for any structure [41]. Data, is converted into a series of consecutive machine tasks that are fed into the Contour Crafting machine. A path with the lowest possible cost for each machine task is determined [41]. It is an on-site process using a crane-mounted press device. The cement paste is extruded through the nozzle against a trowel, allowing the printed element to cover a smooth surface during the process.

**Fig. 5.6** CONPrint3D
technology in contour crafting
construction operation [30]

### 5.2.1.2   Particle Bed Fusion

Particle bed fusion (also called powder bed print, binder jet or sand print) is a production method in which a layer of particles is spread out and, a print head moves and selects the particle and deposits it on the particles as a binder (Fig. 5.7). Then another layer of particles is spread and this process is repeated so that one component can be layered onto the layer. The process is advantageous so that the particle bed functions as a temporary support structure, enabling angling. As part of producing cemented components, this process was recently pioneered by Lowke [42].

Water can be added to a bed of binder, cemented materials, or the binder can be like a paste injected into an aggregate bed in the production of these components [44]. In addition to being able to produce angled or undercut parts, one of the primary advantages of this method is its high resolution. Theoretically, the smallest to the largest particle in the bed is soluble. This method also requires a final operation step to remove these unbound particles, which are generally limited in terms of recyclability in Portland cement systems [45].

- **D-Shape**

D-Shape is an off-field layer printing process invented by Italian inventor Enrico Dini in 2007. First version *D-Shape* binds the sand layers of the printer with inorganic seawater and magnesium-based binder and creates stone-like objects [46]. Current versions of the D-Shape printers arrange any mixed grained material to an 'Ink Binder' fibers with a viscosity between viscosity and water from 0.1 to 4 mm (and theoretically up to 20 mm) diameter. Thanks to thin layers (thickness: 5–

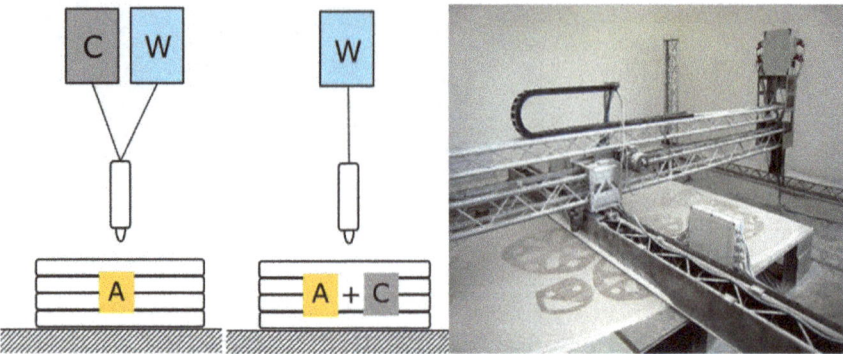

**Fig. 5.7** Schematic of particle bed fusion technology process. C, W, and A; cement, water, and aggregate respectively [33, 43]

10 mm) and a large number of printhead nozzles, this technology provides high resolutions and excellent geometric freedom. Therefore, particle bed 3D printing is especially suitable for the production of geometric complex architectural elements [46]. The newest large-scale pilot project using a *D-Shape* printer is the world's first pedestrian bridge built in Madrid, Spain in 2018 [47]. The bridge was designed at the Catalonia Institute for Advanced Architecture (IAAC) using a parametric design that allows minimizing the amount of material used. The bridge with a total length of 12 m and a width of 1.75 m *D-Shape* printed using printer and micro reinforced concrete [30]. The particle bed technology was developed as part of a research project at the Technical University of Munich under the DFG Priority Program "Leicht Bauen mit Beton" in Germany [42, 48]. The researchers focused on the fundamentals of selective cement activation technology in this project as well as studies on suitable printing materials. Figure 5.8, shows a component manufactured at TU Munich.

**Fig. 5.8** A prototype manufactured at TU Munich

**Fig. 5.9** Similarities between
three types of printing
methods [49]

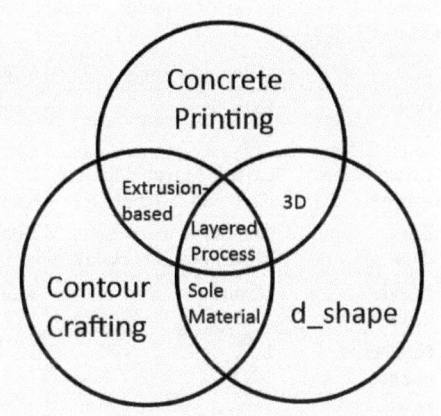

*Contour Crafting* and *Concrete Printing* processes are wet fresh concrete printing processes while *D-Shape* method is a process with dry material. As a result of the process, there is no need for grinding and polishing since smooth surfaces are obtained during the *Contour Crafting* application. But if a smooth surface is desired in *Concrete Printing* and *D-Shape* techniques, grinding, and polishing should be done. However, it is a rough surface to be applied to the element, conventional processes such as plaster, insulation, etc. might be useful. The printing properties of these three techniques are given in Fig. 5.9 [49] and their comparison is given in Table 5.1.

### 5.2.1.3  Spraying

It is an innovative printing technique recently developed by researchers at Braunschweig Technical University in Germany. The method combines the basic principles of well-known shotcrete technology with robotic digital production capabilities [50]. Contrary to conventional multi-layer extrusion, concrete is not extruded in spraying technology, but the layer is applied through spraying by adding controlled compressed air to the extrusion mold (Fig. 5.10). The large-scale printer produced in the Digital Building Production Laboratory in TU Braunschweig consists of a CNC-controlled 5-axis portal machine and a 6-axis heavy-duty robot. The robot arm is equipped with a nozzle charged by an eccentric screw pump. Due to the extended degrees of freedom of the robotic arm (i.e., pressure jet) and the good layer limited by the spraying of the material, the spraying technology enables changing the angle of application in the production process. This allows greater geometric freedom of the formed shapes. In addition to robot-assisted filler manufacturing, this system enables both CNC-controlled finishing with smoothing, milling and, other interactive processes [39].

**Table 5.1** Comparison of contour crafting, concrete printing, and d-shape concrete printing techniques [29]

|  | Contour crafting | Concrete printing | D-shape |
|---|---|---|---|
| Process | Extrusion | Extrusion | Three-dimensional printing |
| Mold usage | Yes | No | No |
| Construction material | Cemented material for mold and building | Printable concrete for homes | Granular material (sand/ gypsum powder) |
| Binder material | None (wet material extrusion and backfill) | None (wet material) | Chlorine-based liquid |
| Nozzle diameter | 15 mm | 9–20 mm | 0.15 mm |
| Number of nozzles | 1 | 1 | 6–300 |
| Layer thickness | 13 mm | 6–25 mm | 4–6 mm |
| Supporting | Yes | Yes | No |
| Mechanical features | Layer orientation was tested at 0 °C. So this force is applied from the top of the printing surface | | |
| Compressive strength | Uncertain | 100–110 MPa | 235–242 MPa |
| Bending strength | Uncertain | 12–13 MPa | 14–19 MPa |
| Print size | >1 m | >1 m | >1 m |
| First and last process | Vertical one per 125 mm reinforcement and backfill with cemented material | Post-printing reinforcement | Light pressure with a roller before the deposition to compact the powder in the next layer, removal of unused material |
| Pros | Smooth surface with a trowel | High strength | High strength, minimum printing process; deposition and reinforcement |
| Cons | Extra molding process Poor bonding between layers due to backfill groups every 1 h | Limited print size due to print frame 5.4 m (length) × 4.4 m (width) × 5.4 m (height) | A slow process, hard surface, limited print size (print frame), large material insertion, removal of unused material |

### 5.2.1.4 Sliding Mold

Slipforming is a method in which a vertically moving mold is placed on top in a liquid state and comes out in a solidified state. The dynamic casting process of slipforming was invented in 1899 by engineer Charles F. Haglin in collaboration with grain company owner Frank Peavey [51]. It is a one-stage process in which the concrete is continuously placed in a constrained mold that moves vertically at a speed adjusted according to the hydration rate of the concrete and allows the

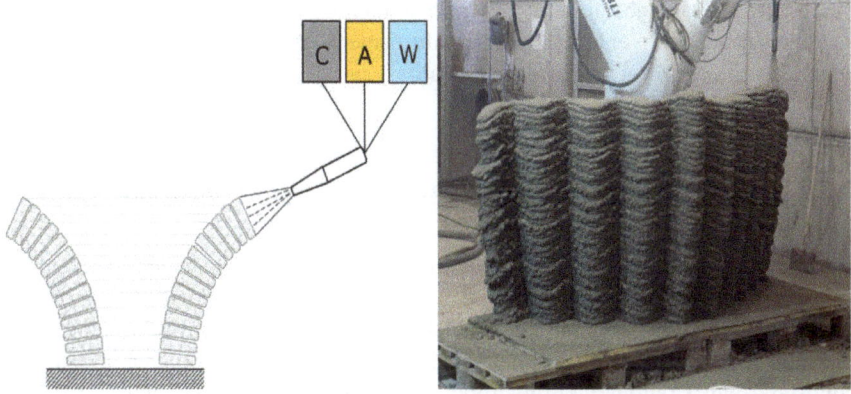

**Fig. 5.10** Scheme of 3D concrete printing process with spraying technology, spraying layers of pre-mixed material. C, W, and A; cement, water, and aggregate respectively [50]

material to support itself when released by the mold (Fig. 5.11). Unlike extrusion, the concrete is not forced from the mold [52]. This system is suitable for producing column elements of variable cross-section using a digitally operated, deformable mold, but it has also been used to construct thinly folded structures that are more resistant to bending [53]. A similar approach has been applied to produce a large array of bent column elements in order to use them in a variable shade façade [54].

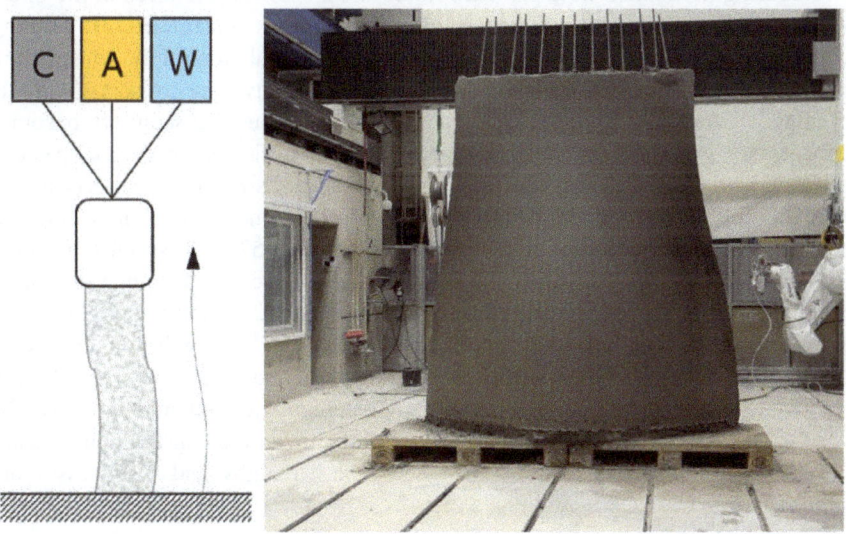

**Fig. 5.11** Scheme of the intelligent dynamic casting process, the pre-mixed material is filled into the vertically moving print head with the speed adjusted to the tempering kinetics of the concrete [33]

The biggest advantages of sliding in comparison to other techniques are the ability to slide around conventional steel reinforcement, the relative absence of layer interface problems, and high surface quality.

## 5.2.2 The Potential and Challenges of 3D Concrete Printing

The growing interest of researchers and the industry, numerous new initiatives and, projects are already an indication that many participants know the significant advantages and improvement potential that additive manufacturing technologies can offer to the concrete industry. CyBe Additive Industries, a Dutch company that builds its own Contour Crafting robot, has produced a special mortar for their robots. It is known that this mortar produces 32% less $CO_2$ than normal concrete and is more environmentally friendly. The mortar is also completely reusable, so waste and environmental pollution can be greatly reduced [55].

CyBe is not the only company that produces a more sustainable mix for its 3D printers, WinSun, a Chinese company with a Contour Crafting printer, uses "a printing material made from recycled construction waste, industrial waste and, other wastes" [56]. Another advantage linked to the sustainability of 3D printed concrete is the reduction of construction waste by 30–60%. In addition to improved sustainability, Michell Starr [57] asserts that there is a "50–70% reduction in production time and 50–80% in labor costs". According to Buswell [29], the reduction in cost and time can be advantageous for both contractors and customers.

Reducing production time provides fewer and shorter disturbances in the direct vicinity of the 3D-printed building or object where it is constructed. 3D printing will also reduce the inconvenience with an unconventional order of the construction site, only the machinery is required—maybe with one or two supervisors—with 3D printing all traditionally used equipment that often causes discomfort becomes unnecessary. A more creative advantage of 3D printing is that specific shapes can be made with ease. This removes the limits for standard sizes that architects should normally follow. Second, 3D printing no longer means mass-producing a standard product to fit all purposes. According to Kolarevic [57], "mass customization technologies and methods allow the creation and production of unique or similar buildings and constructing components differentiated by digitally controlled variations". It seems that future owners of a 3D-printed building will be able to customize their building to their wishes without paying extravagant costs. Scholarship has pointed out that the combination of digital technologies and automation can be very important steps in solving productivity problems in the construction industry [14, 30]. In the analysis shown in Table 5.2, strengths and weaknesses are summarized.

**Table 5.2** Swot analysis [58]

| Strengths | Weaknesses/problem areas |
|---|---|
| • Accelerating product development<br>• A decrease in cycle time from design to manufacturing<br>• Easy manufacture of intricate parts<br>• Increasing demand for product designers<br>• Educational institutions using 3D printers to accelerate learning and understanding<br>• The democratization of creative/ manufacturing power<br>• Traditional mass production responding to the threat by finding ways to reduce costs and minimum production quantities<br>• An increasing number of open-source "plans" that can be used for printing an array of objects<br>• Providing environmental benefits by reducing transportation needs | • An increased amount of waste and the burden on the environment<br>• The production of non-isotropic parts in the layer process, i.e., parts whose strengths are not equal in all directions, limits the functionality of the parts<br>• Workplace losses in deteriorating sectors<br>• The prominence of intellectual property as a value source in productivity<br>• Piracy<br>• Brand and product quality<br>• High maintenance cost<br>• Low speed<br>• High investment cost<br>• Limited product sizes, not allowing scale production |
| The impact on the construction industry | |
| Opportunities | Threats |
| • The emergence of new venture opportunities in space<br>• The emergence of a new industry supplying printing materials<br>• Increasing job opportunities with fast production<br>• Contribution to local activities | • The potential for any innovation to be copied instantly<br>• The emergence of the opportunity to 3D printing products with potential malevolent functions<br>• Increasing waste, increasing the burden on the environment |

## 5.3 Bibliometric Analysis

Systematic reviews of the scientific literature in a particular field of research are frequently encountered thanks to the development of tools that can perform long-term analyzes, which require a long time to obtain with traditional methods, using automatic, effective and, fast techniques, and the increased availability of online databases [59].

The descriptive, social and intellectual structure of the bibliometrically analyzed data was examined in this study. Open-source code R bibliometrix package and R software were used for analysis. 4208 articles examining the subjects of "3D" and "concrete" or "digital" and "concrete" in the WOS database were analyzed by limiting the starting year from 2000 until present within the scope of the research. The parameter of *(TITLE-ABS-KEY ("3D" AND "concrete" OR "digital" AND "concrete")) AND (LIMIT-TO (YEAR, "2000–2020")) AND (LIMIT-TO (LANGUAGE, "English"))* has been determined while searching the WOS database. With the help of this search parameter, the studies involving "3D" AND "concrete" OR "digital" AND "concrete" in the title, abstract and keywords are listed. After the data set was uploaded into the R statistical program, descriptive analysis, network analysis and, cluster analysis on networks were performed. In addition to co-occurrence analyzes, co-citation and descriptive data analyzes were performed on the obtained data. Since there is no consensus on it, which of the databases such as WOS, Google Scholar, Scopus to choose is usually up to the researchers' preference. The WOS database has been preferred because it has proven itself in the field of architecture and engineering and has reliable sources.

There are a total of 11,849 different keywords in studies published on "3D and concrete/digital and concrete" in 961 different sources (such as journals, conferences) in the fields of architecture and engineering works for approximately 20 years (2000–2020). These publications were cited an average of 12.66 times. It was observed that 10,426 authors contributed to these studies and the names of the authors appeared in publications 15,687 times. It is seen that a significant portion of the publications (97.88%) have more than one author and the number of authors per article is 2.48. *Collaboration Index* (CI) is another index value based on the authors' collaboration. This index value deals with multi-author publications in the field and is calculated as *authors of multi-authored doc/multi-authored doc* (Ajiferuke, Burell and Tague, 1988). Thus, the effect of single-author publications is nulled and the number of authors per publication is observed in publications with multiple authors. In this context, the CI value of the printable concrete works made in the fields of architecture and engineering was determined as 2.57 (Table 5.3).

It is seen that printable concrete studies were introduced in the field of architecture and engineering in 1976. But a time limitation has been imposed to include studies from 2000 to the present. In this context, it is observed that there is a rapid

**Table 5.3** General information

| Description | Results |
| --- | --- |
| Documents | 4208 |
| Documents per year | 4.85 |
| Sources (Journals, Books, etc.) | 961 |
| Keywords | 11,849 |
| Years | 2000–2020 |
| Average citations per documents | 12.66 |
| Average citations per year per documents | 2 |
| References | 103,039 |
| Authors | 10,426 |
| Author appearances | 15,687 |
| Authors of single-authored documents | 221 |
| Authors of multi-authored documents | 10,205 |
| Single-authored documents | 237 |
| Documents per author | 0.404 |
| Authors per document | 2.48 |
| Co-authors per documents | 3.73 |
| Collaboration index | 2.57 |

**Table 5.4** Publication numbers by years

| Year | 2000 | 2001 | 2002 | 2003 | 2004 | 2005 | 2006 | 2007 | 2008 | 2009 | 2010 |
|---|---|---|---|---|---|---|---|---|---|---|---|
| Article | 28 | 38 | 31 | 41 | 46 | 33 | 65 | 59 | 88 | 104 | 121 |
| Year | 2011 | 2012 | 2013 | 2014 | 2015 | 2016 | 2017 | 2018 | 2019 | 2020 | |
| Article | 144 | 159 | 192 | 213 | 281 | 363 | 468 | 602 | 684 | 404 | |

increase in the publications that examine the growth data of the 20-year research history (Table 5.4) and an average of 4.85 publications per year are included in the literature (Table 5.3). As observable in Fig. 5.12, it was determined that the number of publications made by years has increased consistently since 2007.

4208 articles containing the concepts of "3D" and "concrete" or "digital" and "concrete" were found in the WOS database for the field of architecture and engineering. When the keywords of these articles are examined "*concrete*", "*digital image correlation*", "*3d printing*", "*reinforced concrete*", "*digital image*", "*fracture*" are the most frequently observed concepts (Fig. 5.13).

Articles containing the subject of printable concrete and studies included in the bibliography of these articles (articles, papers, books, etc.) are compared in Table 5.5. When the articles that stand out in this context are examined, it is observed that Ozbolt's articles are cited in significant numbers in the bibliography of the journals, while Ozbolt's articles were not found in the list of articles that received a significant number of citations in journals.

The Three-Fields Diagram was used to provide an overview of the publications covering the concepts of "3D" and "concrete" or "digital" and "concrete" in the field of architecture and engineering for the years between 2000 and 2020. This

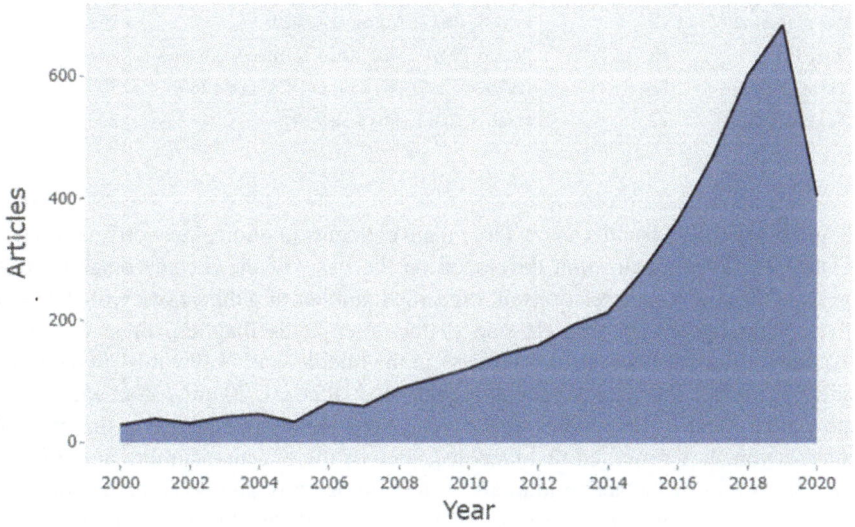

**Fig. 5.12** Publication numbers by years

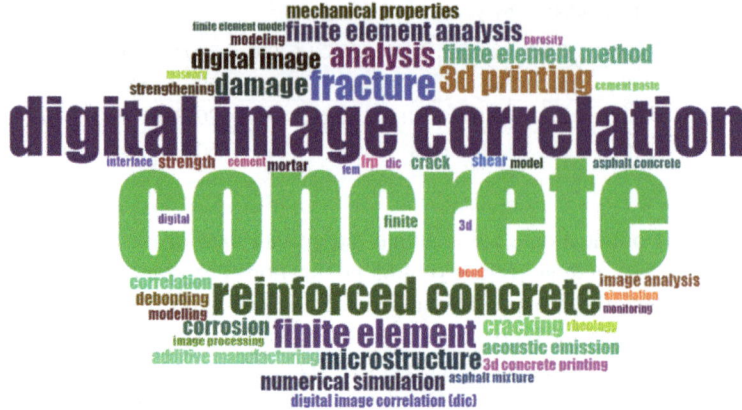

**Fig. 5.13** Keyword cloud

**Table 5.5** Number of citations for articles

| Most cited authors | Local citation | Most cited articles | Global citation |
|---|---|---|---|
| Ozbolt J | 30 | Pointcheval D, 2000, J Cryptol | 982 |
| Tan MJ | 28 | Leung T, 2001, Int J Comput Vision | 865 |
| Wang L | 24 | Bai YL, 2008, Int J Plasticity | 676 |
| Zhang J | 23 | Stansbury JW, 2016, Dent Mater | 358 |
| Sun W | 21 | Lilliu G, 2003, Eng Fract Mech | 268 |
| Liu Y | 20 | Boylan-Kolchin M, 2008, Mon Not R Astron Soc | 263 |
| Mechtcherine V | 19 | Lim S, 2012, Automat Constr | 251 |
| Ding FX | 18 | Areias PMA, 2005, Int J Numer Meth Eng | 244 |
| Varma AH | 18 | Gallucci E, 2007, Cement Concrete Res | 207 |
| Aggelis DG | 17 | Haah J, 2011, Phys Rev A | 203 |

diagram basically visualizes how three main elements (authors, keywords, journals) relate [60]. The visualization developed on the basis of the Sankey diagram [61] presents the component relations of a specified number of publications with the help of a single figure. On the left side of the three-fields diagram, there are cited "sources" (journal, book, etc.), "authors" in the middle, and "keywords" in the right side (Fig. 5.14). The analysis involving the most effective 20 units from each field, shows that most of the studies directly cite the journals publishing in the field of construction. It is observed that Ozbolt J., one of the effective authors in the field, cites almost all of the active journals in the bibliography and he is the author who uses the word "*concrete*" the most. It is understood that the publications of the

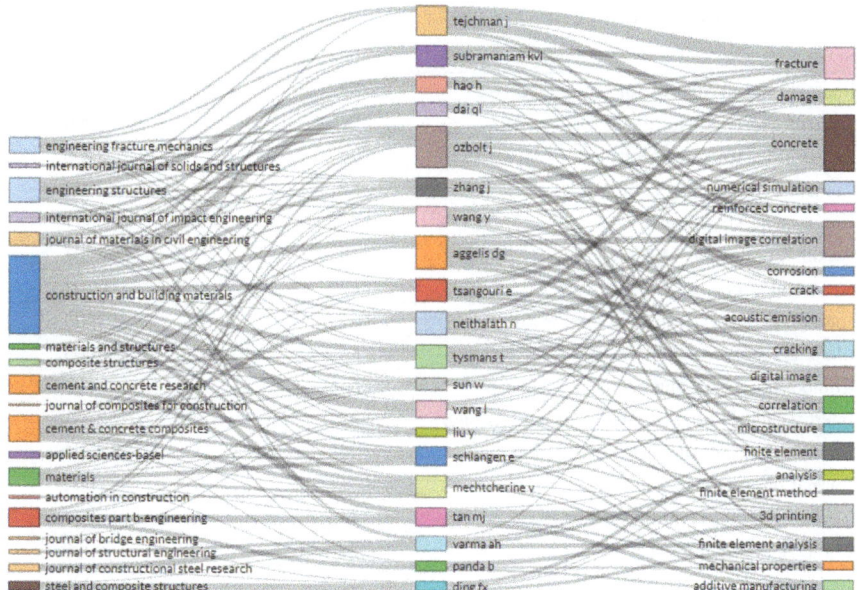

**Fig. 5.14** Three-fields diagram

influential authors in the three field diagrams such as Tejchman J., Aggelis D.G. and, Tsangorui E., received significant citations. It is seen that the words *concrete*, *digital image* and *correlation* come to the fore in keywords preferred by effective research in the field.

The journal with the most published works in the field of architecture and engineering is Construction and Building Materials. It is seen that the journal started publishing studies in this field in 2004. The journal is the source for publishing key studies in the field (Table 5.6). It is found that although the "*Materials*" journal started to publish in 2016, it has become an important source in this field. Keyword trend analysis shows the patterns in the literature over the years, on a coordinate plane, by using the keywords, abstracts and, titles in the research. It is pointed out which subjects used the studies published in which years more in this analysis. There are certain benefits of correct keyword selection, title and, abstract writing in scientific publications. Preliminary definitions such as abstract, title and, keyword in scientific publications are important in terms of describing which research field the publications are about and its framework. Thus, it is easier for other researchers to access publications belonging to the relevant literature, thanks to the framework drawn by keywords, titles and articles [62].

Thanks to the trend topics analysis, it is possible to observe that the keywords, titles and, abstracts, which have become a driving factor for the widespread use of research in any field, have changed over the years (Table 5.6). Keyword trend analysis assigns terms to the coordinate plane according to logarithmic frequency values to crytallize this change. The current situation of printable concrete works in

**Table 5.6** Number of articles in relevant journals

| Sources | Number of articles | Year |
|---|---|---|
| International journal of impact engineering | 33 | 2000 |
| Cement and concrete research | 91 | 2000 |
| Engineering fracture mechanics | 79 | 2000 |
| Cement and concrete composites | 85 | 2001 |
| Engineering structures | 223 | 2001 |
| Automation in construction | 49 | 2001 |
| Structural concrete | 32 | 2003 |
| Journal of materials in civil engineering | 65 | 2003 |
| Materials and structures | 64 | 2003 |
| Composites part b-engineering | 53 | 2003 |
| International journal of pavement engineering | 33 | 2003 |
| Journal of composites for construction | 35 | 2004 |
| Journal of constructional steel research | 35 | 2004 |
| Construction and building materials | 445 | 2004 |
| Composite structures | 58 | 2006 |
| Journal of structural engineering | 42 | 2006 |
| Journal of bridge engineering | 28 | 2009 |
| International journal of solids and structures | 31 | 2013 |
| Applied sciences-basel | 40 | 2015 |
| Materials | 57 | 2016 |

the field of architecture and engineering was examined with the help of Fig. 5.15, which was created to show 3 topics (obtained from abstract, keyword and title) that examined at least 3 times a year between 2000 and 2020. The trends show that as the research dates get closer to today, the studies include the keywords of the 3D printers and concrete printers.

Collaboration networks present joint publications by individuals or organizations such as countries, authors, or journals (organizers) through a social network. In the cooperation network by country, the adjacency matrix *country x country adjacency matrix* which is basically determined according to the frequency of publishing together is used. Network analysis conducted according to co-authoring status is specified in Formula 5.1 with modularity score $Q$ [63].

$$Q = \frac{1}{2h} \sum_{ii'} \left[ a_{ii'} - \frac{\delta_i \delta_{i'}}{2h} \right] s_i s_{i'} \qquad (5.1)$$

**Formula** 5.1. Network analysis modularity score formula

$\delta_i$, $i$ represents the degree of node, as $h$ represents the total number of ties in the network and $S_i$, $i$ represents the node's membership to a community in the

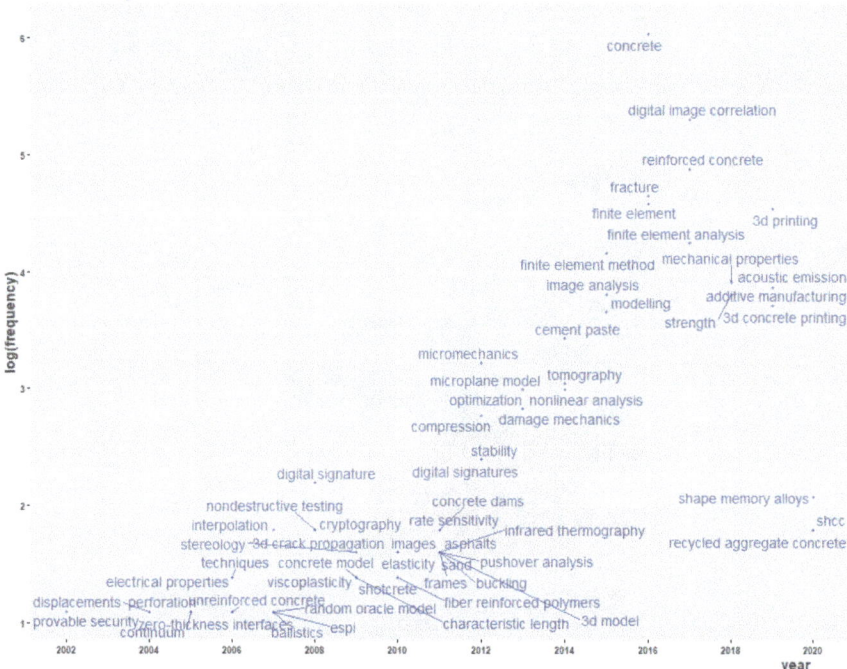

**Fig. 5.15** Logarithmic frequencies of words by years

algorithm. In the network where collaborations by university are examined according to the frequency of publishing together, *university x university* and the adjacency matrix (country *x* country adjacency matrix) is used.

Figure 5.16 shows that several countries cooperate in printable concrete studies in the field of architecture and engineering. The size of the nodes in the network, colors of the countries' publishing frequency provides the information about the clusters they belong to. It is observed in Fig. 5.16 that clusters in the same color tend to publish together more. Accordingly, the UK, China, Australia, USA, UK and, Turkey form the red cluster; Malaysia, Canada, Arabia and, Iran green cluster; The Netherlands, Italy, Portugal, Denmark, New Zealand, Germany and, Switzerland the blue cluster, respectively. The relationship in the network, which includes a total of 37 countries, emerges from the clustering of nodes with at least two ties. The growth of ties indicates the greater number of relations between nodes and the growth of nodes indicates the magnitude of the activity of a node in the network.

When the collaboration of the universities or research centers affiliated with the authors who make printable concrete studies in the field of architecture and engineering is examined (Fig. 5.17), "Beijing University" in the red cluster, "Southeast University" in the blue cluster, "Delft University" in the purple cluster, "Tongji University" in the green cluster, "Dalian University" in the orange cluster and,

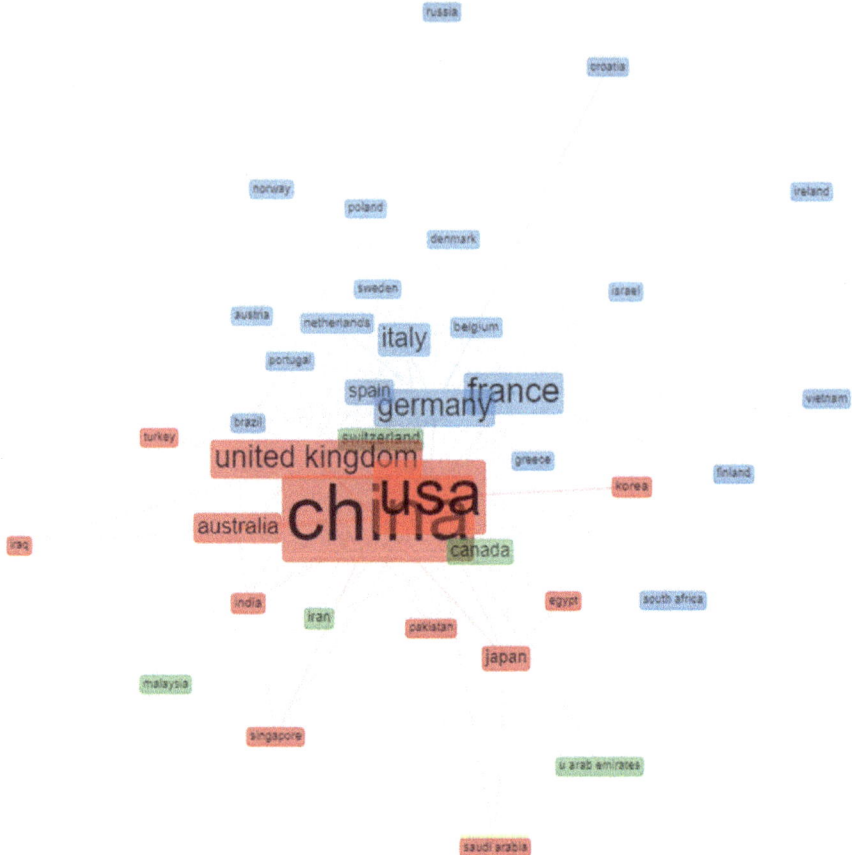

**Fig. 5.16** Country collaboration network

"Cents University" stands out in the brown cluster, respectively. Due to the social structure of the network, a map of nodes with two or more relations could not be formed. Instead, 29 universities with nodes with at least one link and that meet the criteria were represented in the social network.

For the collaboration of authors who published printable concrete works in the field of architecture and engineering, the cases of two documents' (article, etc.) inclusion together in other documents are analyzed. The more often these documents are included in other documents, the larger the nodes in the collaboration network will grow and become closer to each other. The cooperation between the authors of the articles in the journals in the field of architecture and engineering is visualized in Fig. 5.18. It is found that the authors, who have important positions on the network, cluster in blue, red, brown, green, grey, orange and, purple colors. Authors represented by the same color and close knots appear together more frequently in the bibliography of other documents.

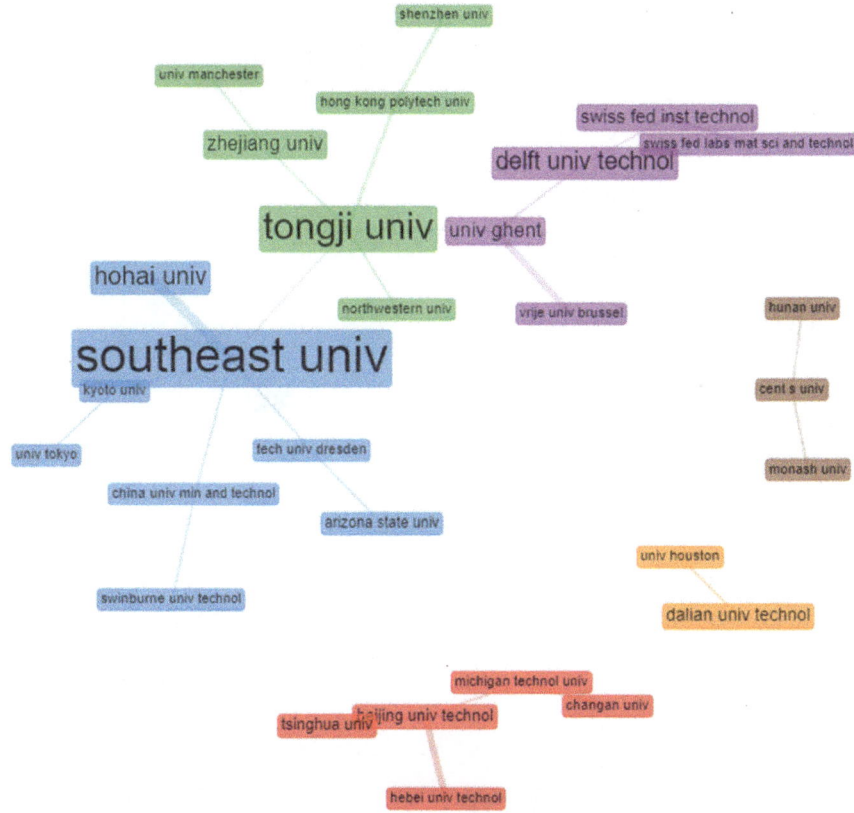

**Fig. 5.17** University collaboration network

## 5.4   Evaluation of Findings and Conclusions

Bibliometric analyzes, which are becoming increasingly substantial for scientific disciplines, enable researchers to access data in a scientific field with a panoramic perspective. As the number of publications continues to increase, it becomes increasingly difficult to turn the data obtained from these publications into useful information using traditional methods. Determining the intellectual structure and research front of scientific fields is not only significant for new research but also sectoral and public applications. One of the software that has the ability to analyze bibliometry was utilized in this study. The R program offers useful solutions to its users with its open-source libraries that have extensive support and the possibility to make many analyzes. This study involves some suggestions for other studies. Publications and articles in the WoS database in English were discussed within this study. Examining different academic databases (Scopus, etc.) and different types of studies (thesis, book, etc.) of other studies has the potential to open new horizons to

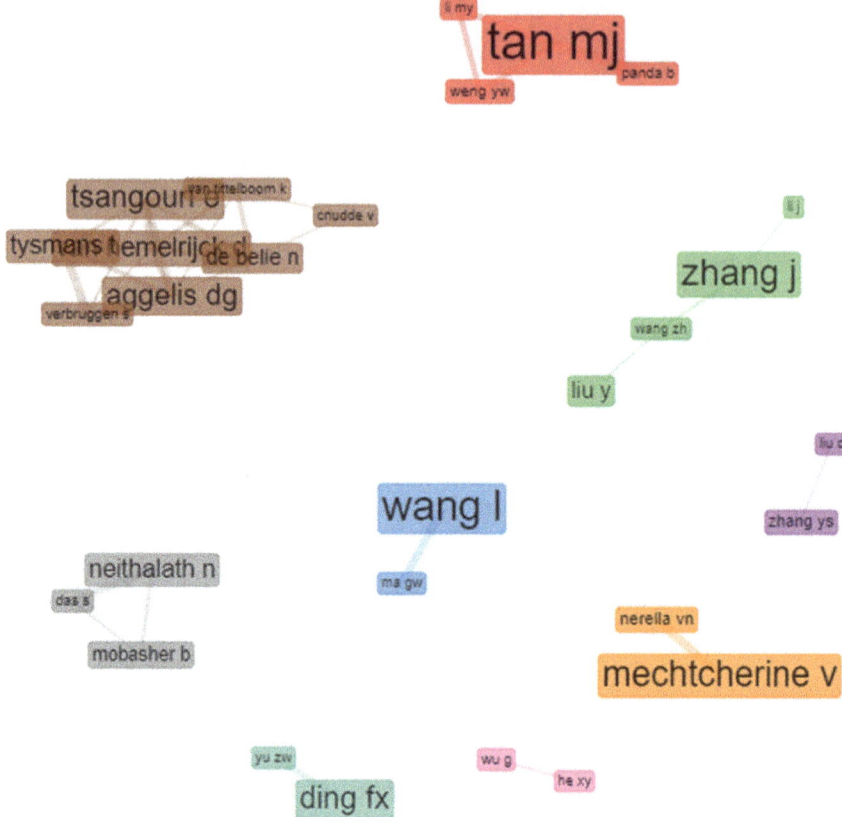

**Fig. 5.18** Author collaboration network

research. The research contains some limitations in itself. The preference of English articles in the Wos database during the acquisition of the reviewed articles results in ignoring the researches made in other languages. In addition, findings of the studies scanned outside the WoS database were not included. It is thought that different bibliometric techniques (co-citation, theme analysis, etc.) to be used in future research will contribute to the growth of the research. This study presents a panoramic perspective of the literature. In this context, the results of the research are expected to provide descriptive, social and, intellectual information about the subject to academicians and graduate students who want to do new research in the field of printable concrete in the disciplines of architecture and engineering. 4208 articles were examined within the scope of the research and noteworthy findings were obtained about the concept of printable concrete. The factorial analysis results obtained in the research findings indicate that the clusters contain different approaches for architecture and engineering. It is seen in this study that the institutions and countries that are geographically close to each other tend to publish together more in the cooperation networks formed by considering the countries of the authors and the universities they are affiliated with. Nevertheless, the collaboration of authors who publish from countries such as America, England, Australia and, China, which are frequently researched, with the participants of many other countries increased the effectiveness of these countries in the network. Another noteworthy point is that Turkey is 23rd among the countries that publish on the subject while the citation rate is low compared to countries that publish less than it. This study is thought to provide a base for the publications to be conducted in Turkey. This research is valuable in terms of providing an idea about the structure of the subject to researchers who will work on printable concrete. Researchers can gain insights by observing current research topics in printable concrete from a broad perspective, such as smart cities. Bibliometry provides important benefits in terms of revealing the topics studied in the relevant literature, trends and potential research topics.

# References

1. World Economic Forum.: Shaping the Future of Construction—A Breakthrough in Mindset and Technology, Geneva, Switzerland. http://w3.weforum.org/docs/WEF_Shaping_the_Future_of_Construction_full_report__.pdf. Accessed 8 Aug 2018 (2018)
2. Global Construction Perspectives and Oxford Economics.: A Global Forecast for the Construction Industry by 2030, London, UK. ISBN 978-0-9564207-9-4 (2015)
3. European Commission.: The European Construction Sector: A Global Partner (2016). https://ec.europa.eu/docsroom/documents/15866/attachments/1/translations/en/renditis
4. Dalenogare, L.S., Benitez, G.B., Ayala, N.F., Frank, A.G.: The expected contribution of Industry 4.0 technologies for industrial performance. Int. J. Prod. Econ. **204**, 383–394 (2018). https://doi.org/10.1016/j.ijpe.2018.08.019
5. Stock, T., Seliger, G.: Opportunities of sustainable manufacturing in Industry 4.0. Procedia CIRP **40**, 536–541 (2016). https://doi.org/10.1016/j.procir.2016.01.129
6. Gerbert, P., Castagnino, S., Rothballer, S., Renz, A., Filitz, R.: Digital in Engineering and Construction. The Boston Consulting Group, USA (2016). http://futureofconstruction.org/content/uploads/2016/09/BCG-Digital-in-Engineering-andConstruction-Mar-2016.pdf
7. Craveiro, F., Duarte, J.P., Bartolo, H., Bartolo, P.J.: Additive manufacturing as an enabling technology for digital construction: a perspective on construction 4.0. Autom. Constr. **103**, 251–267 (2019)
8. Al-Qutaifi, S., Nazari, A., ve Bagheri, A.: Mechanical properties of layered geopolymer structures applicable in concrete 3D-printing. Constr. Build. Mater. J. **176**, 690–699 (2018). https://doi.org/10.1016/j.conbuildmat.2018.04.195
9. Håkansson, H., Ingemansson, M.: Industrial renewal within the construction network. Constr. Manag. Econ. **31**, 40–61 (2013). https://doi.org/10.1080/01446193.2012.737470
10. Dilberoglu, U.M., Gharehpapagh, B., Yaman, U., Dolen, M.: The role of additive manufacturing in the era of Industry 4.0. In: 27th International Conference on Flexible Automation and Intelligent Manufacturing, FAIM2017, Italy (2017)
11. European Construction Industry Federation.: Construction 4.0: Challenges and Opportunities (2017). http://www.fiec.eu/en/themes-72/construction-40.aspx
12. García de Soto, B., Agustí-Juan, I., Hunhevicz, J., Joss, S., Graser, K., Habert, G., Adey, B.T.: Productivity of digital fabrication in construction: cost and time analysis of a robotically built wall. Autom. Constr. **92**, 297–311 (2018). https://doi.org/10.1016/j.autcon.2018.04.004

13. Ghaffar, S.H., Corker, J., Fan, M.: Additive manufacturing technology and its implementation in construction as an eco-innovative solution. Autom. Constr. **93**, 1–11 (2018). https://doi.org/10.1016/j.autcon.2018.05.005
14. Wangler, T., Roussel, N., Bos, F.P., Salet, T.A.M., Flatt, R.J.: Digital concrete: a review. Cem. Concr. Res. **123**, 105 (2019)
15. Hall, C.M.: Publish and perish? Bibliometric analysis, journal ranking and the assessment of research quality in tourism. Tourism Manage **32**(1), 16–27 (2011)
16. Chen, C., San Juan, F., Hou, J.: The structure and dynamics of cocitation clusters: a multiple perspective cocitation analysis. J. Am. Soc. Inf. Sci. Technol. **61**(7), 1386–1409 (2010)
17. Van Eck, N., Waltman, L.: Software survey: VOSviewer, a computer program for bibliometric mapping. Scientometrics **84**(2), 523–538 (2010)
18. Bos, F., Wolfs, R., Ahmed, Z., ve Salet, T.: Additive manufacturing of concrete in construction: potentials and challenges of 3D concrete printing. Virtual Phys. Prototyp. **11**, 209–225 (2016). https://doi.org/10.1080/17452759.2016.1209867
19. De Schutter, G., Lesage, K., Mechtcherine, V., Nerella, V.N., Habert, G., Agusti-Juan, I.: Vision of 3D printing with concrete-technical, economic and environmental potentials. Cem. Concr. Res. **112**, 25–36 (2018). https://doi.org/10.1016/j.cemconres.2018.06.001
20. Bulut, F.E.: An Assessment of the Potential and Future of 3D Printer Technologies in Architecture: A Social, Economic and Environmental Perspective. Master Thesis, Gazi University Institute of Science and Technology, Ankara (2019)
21. Ashby, M.: Materials Selection in Mechanical Design, 4th edn (2010). https://www.elsevier.com/books/materials-selection-in-mechanical-design/ashby/978-0-08-095223-9. Accessed 20 Jan 2019
22. Sakin, M., Kiroglu, Y.C.: 3D Printing of buildings: construction of the sustainable houses of the future by BIM. In: 9th International Conference on Sustainability in Energy and Buildings, SEB 17, 5–7 July 2017, Chania, Crete, Greece (2017)
23. Internet Access.: URL-1, Office of the Future. https://www.asme.org/engineering-topics/articles/manufacturing-design/3d-printed-office-the-future. Accessed 01 Oct 2020
24. Internet Access.: URL-2, Office of the Future. https://www.archdaily.com/875642/office-of-the-future-killa-design. Accessed 20 Nov 2020
25. Arslan Selçuk, S., Gönenç Sorguç, A.: From computer screen to construction site. Yapı Dergisi **407**, 154–160 (2015)
26. Internet Access.: URL-2. https://www.archdaily.com/875642/office-of-the-future-killa-design. Accessed 20 Nov 2020
27. Internet Access.: URL-3, Winsun. https://www.archdaily.com/591331/chinese-company-creates-the-world-s-tallest-3d-printed-building. Accessed 20 Nov 2020
28. Internet Access.: URL-4. https://www.archdaily.com/591331/chinese-company-creates-the-world-s-tallest-3d-printed-building. Accessed 01 Oct 2020
29. Buswell, R.: Digital fabrication in the concrete industry. In: Presentation at the 7th RILEM 276-TC Meeting, September 2018, Zurich (2018)
30. Vasilic, K.: Additive manufacturing with concrete. In: German Society for Concrete and Construction Technology (DBV), Berlin, Germany (2020)
31. Gosselin, C., Duballet, R., Roux, P., Gaudillière, N., Dirrenberger, J., Morel, P.: Largescale 3D printing of ultra-high performance concrete—a new processing route for architects and builders. Mater. Des. **100**, 102–109 (2016). https://doi.org/10.1016/j.matdes.2016.03.097
32. Internet Access.: URL-5, Concrete Beam Created with 3D Printing—Wasp. https://www.3dwasp.com/en/concrete-beam-created-with-3d-printing/. Accessed 01 Oct 2020
33. Henke, K.: Additive Fertigung im Bauwesen—Verfahren der additiven Baufertigung, Potentiale von Holzbeton und anderen Leichtbaustoffen, presented at the 19. Augsburger Seminar für additive Fertigung—Prozessketten und digitale Werkzeuge, München (2015)
34. Lim, S., Le, T., Webster, J., Buswell, R., Austin, A., Gibb, A., Thorpe, T.: Fabricating construction components using layered manufacturing technology. In: Presented at the Global Innovation in Construction Conference 2009 (GICC'09), Loughborough University, Leicestershire, UK, 13–16 September, 2009 (2009)

35. Internet Access.: URL-6, 3D Concrete Printing. 3D Concrete Printing|Loughborough University. http://www.lboro.ac.uk/enterprise/3dcp/
36. Salet, T.A.M., Ahmed, Z.Y., Bos, F.P., Laagland, H.L.M.: Design of a 3D printed concrete bridge by testing. Virtual Phys. Prototyping **3**, 222–236 (2018)
37. Khoshnevis, B.: Innovative rapid prototyping process making large sized, smooth surface complex shapes in a wide variety of materials. Mater. Technol. **13**(52–63), 1998 (1998)
38. Khoshnevis, B.: Automated construction by contour crafting—related robotics and information technologies. Autom. Constr. **13**, 5–19 (2004). https://doi.org/10.1016/j.autcon.2003.08.012
39. Khoshnevis, B., Hwang, D., Yao, K.T., Yeh, Z.: Mega-scale fabrication by contour crafting. Int. J. Ind. Syst. Eng. **1**, 301–320 (2006). https://doi.org/10.1504/IJISE.2006.009791
40. Zhang, J., Khoshnevis, B.: Optimal machine operation planning for construction by contour crafting. Autom. Constr. **29**, 50–67 (2013)
41. Szilvśi-Nagy, M., Matyasi, G.Y.: Analysis of STL files. Math. Comput. Model. **7**, 945–960 (2003)
42. Lowke, D., Weger, D., Henke, K., Talke, D., Winter, S., Gehlen, C.: 3D Drucken von Betonbauteilen durch selektives Binden mit calciumsilikatbasierten Zementen—Erste Ergebnisse zu beton-technologischen und verfahrenstechnischen Einflüssen. In: Proceedings of the 19th International Conference on Building Materials Ibausil, Weimar (2015)
43. Edil Tecnico.: Tecnologie digitali di stampa 3D: l'applicazione in edilizia. https://www.ediltecnico.it/37001/tecnologie-digitalistampa-3d-lapplicazione-in-edilizia/. Accessed 24 Jun 2019 (2016)
44. Pierre, A., Weger, D., Perrot, A., Lowke, D.: Penetration of cement pastes into sand packings during 3D printing: analytical and experimental study. Mater. Struct. **51**(1), 22 (2018)
45. Xia, M., Sanjayan, J.: Method of formulating geopolymer for 3D printing for construction applications. Mater. Des. **110**, 382–390 (2016)
46. Dini, E.: D-Shape Company, Monoblock Three-Dimensional. https://dshape.com/company/. Accessed 20 Nov 2020 (2004)
47. Internet Access.: URL-7, 3D printed bridge—IAAC. https://iaac.net/project/3d-printed-bridge/. Accessed 20 Nov 2020
48. Weger, D., Lowke, D., Gehlen, C.: 3D printing of concrete structures using the selective binding method—effect of concrete technology on contour precision and compressive strength. In: Proceedings of the 11th International PhD Symposium in Civil Engineering, The University of Tokyo, Tokyo (2016)
49. Buswell, R.A., Leal de Silva, W.R., Jones, S.Z., Dirrenberger, J.: 3D printing using concrete extrusion: a roadmap for research. Cem. Concr. Res. **112**, 37–49 (2018)
50. Kloft, H., Hack, N., Lindemann, H.: Shotcrete 3D Printing (SC3DP)—3D-Drucken von großformatigen Betonbauteilen, Bautechnik, 02/19, 54–57 (2019)
51. Lloret, E., Shahab, A.R., Linus, M., Flatt, R.J., Gramazio, F., Kohler, M., Langenberg, S.: Complex concrete structures: merging existing casting techniques with digital fabrication. Comput. Aided Des. **60**, 40–49 (2015). https://doi.org/10.1016/j.cad.2014.02.011
52. Lloret, E., Fritschi, L., Reiter, T., Wangler, F., Gramazio, M., Kohler, R.J.: Flatt, smart dynamic casting: slipforming with flexible formwork—inline measurement and control. HPCCIC Tromsø 2017, Norwegian Concrete Association, 2017. https://doi.org/10.3929/ethz-b-000219663. Paper no. 27 (2017)
53. Szabo, A., Reiter, L., Lloret-Fritschi, E., Gramazio, F., Kohler, M., Flatt, R.J.: Adapting Smart Dynamic Casting to Thin Folded Geometries, pp. 81–93. Springer Int Publishing (2019)
54. Yu, L., Luo, D., Xu, W.: Dynamic Robotic Slip-Form Casting and Eco-Friendly Building Façade Design, pp. 421–433. Springer Int Publishing (2019)
55. Marijnissen, M.P., van der Zee, A.: 3D Concrete Printing in Architecture—A Research on the Potential Benefits of 3D Concrete Printing in Architecture (2017)
56. Starr, M.: World's First 3D-Printed Apartment Building Constructed in China. Retrieved from http://www.cnet.com (2015)
57. Kolarevic, B.: Architecture in the digital age: design and manufacturing. Spon Press, London (2003). https://doi.org/10.1007/s00004-004-0025-4

58. Schwab, K.: The Fourth Industrial Revolution. Crown Publishing Group, New York (2017)
59. Aria, M., Misuraca, M., ve Spano, M.: Mapping the evolution of social research and data science on 30 years of social indicators research. Soc. Ind. Res. 1–29 (2020)
60. Internet Access.: URL-8, Package 'bibliometrix'. https://cran.r-project.org/web/packages/bibliometrix/bibliometrix.pdf. Accessed 01 Oct 2020
61. Riehmann, P., Hanfler, M., Froehlich, B.: Interactive sankey diagrams. In: IEEE Symposium on Information Visualization, INFOVIS 2005 (2005)
62. Grant, M.J.: Key words and their role in information retrieval. Health Inf. Libr. J. **27**(3), 173–175 (2010)
63. Blondel, V.D., Guillaume, J.L., Lambiotte, R., Lefebvre, E.: Fast unfolding of communities in large networks. J. Stat. Mech. Theory Exp. **10**, 108 (2008)

# Chapter 6
# A Material-Driven Design Approach Methodology in 3D Printing Waste Recycling

**Letícia Faria Teixeira, Juliana de Vilhena Rodrigues, Lauro Arthur Farias Paiva Cohen, and Nubia Suely Silva Santos**

**Abstract** This work aims to show 3D printing waste recycling efforts and use them in the design of composites with lignocellulosic fibers as reinforcement. The Material-Driven Design (MDD) approach is a methodology that has been used to bring new meanings for the study of materials, and in this work, it was used in the search for sensory and interpretative aspects of the waste and of the developed samples. The waste used in this work was obtained from 3D printers of Technology Incubators Network with Poly (lactic acid) and Acrylonitrile Butadiene Styrene polymeric filaments. As reinforcement was used Açaí (Euterpe Oleracea Mart.) fiber, Jute (Corchorus capsularis) fiber, and wood flour, which have been added in specific quantities to the polymer during the casting process and then poured in silicone molds. The analyzes were made with the residues and with the samples developed based on the MDD method, when groups of students were formed to interact with the samples, seeking to understand the different factors that influence the user experience with the material under study, finding sensory characteristics that can add new meanings and attributes to the recycled material. As a result, there are different perceptions of the material under study that contribute to the design process, generating products and proposals that are environmentally sustainable and have different meanings for users.

**Keywords** Sustainable design · Circular economy · Polymeric composites · Lignocellulose fibers

**Abbreviations**

**MDD**  Material-Driven Design
**PLA**  Poly (lactic acid)
**ABS**  Acrylonitrile butadiene styrene

L. F. Teixeira · J. de Vilhena Rodrigues · L. A. F. P. Cohen
Bachelor of Design, University of State of Pará, Belem, Brazil

N. S. S. Santos (✉)
University of State of Pará, Belem, Brazil
e-mail: nubiasantos@uepa.br

© The Author(s), under exclusive license to Springer Nature Switzerland AG 2022    105
K. Sandhu et al. (eds.), *Sustainability for 3D Printing*, Springer Tracts
in Additive Manufacturing, https://doi.org/10.1007/978-3-030-75235-4_6

**PETG**  Poly (ethylene terephthalate glycol)
**PA**    Polyamide
**DIY**   Do it yourself
**AM**    Additive manufacturing
**SLA**   Stereolithography
**SLS**   Selective laser sintering
**BJ**    Binder jetting
**FDM**   Fusion deposition modeling
**PP**    Polypropylene
**PC**    Polycarbonate

## 6.1  Introduction

The additive manufacturing technologies are known to generate less waste when compared to conventional subtractive manufacturing such as machining, which is favorable to environmental sustainability. The 3D printing is an extrusion-based technique and is the most used technology due to its low cost is used extensively in teaching and research institutions and technology and innovation centers that take advantage of its versatility as the freedom to manufacture objects in complex shapes and product customization [1]. 3D printers are based on fusion deposition modeling of polymers such as poly (lactic acid) (PLA), Acrylonitrile Butadiene Styrene (ABS), poly (ethylene terephthalate glycol) (PETG), and polyamide (PA). The PLA is a biodegradable polymer that comes from renewable resources, and ABS, a synthetic polymer, is the most used polymers in 3D printers [2, 3]. Nowadays with environmental concerns at the top of the world agenda, the sustainable production and consumption should also guide product development activities in laboratories and maker spaces of teaching and research institutions. Thermoplastic recycling has many benefits to the environment as reducing the amount of plastic in landfills and the need to use virgin resins. In the case of 3D printers, the objects manufactured are of a single material which facilitates recycling after the product disposal, but it also facilitates the process of recycling 3D printer waste generate during printing [4]. The MDD methodology is relevant in this context of waste valuation and seeking new meanings from the experience with users [5]. In contrast to traditional methodologies in which the material selection step is part of the final of the design process, in the MDD method the material (or waste) is the conductor of all the process [6]. Previous studies conducted by this research group have addressed post-consumer plastic recycling, and the development of materials with agro-waste fibers [7, 8].

In this opportunity, the joint work between the Materials and Design Laboratory and Concept Laboratory in Engineering, Processes, and Technology, considers the waste from 3D printers in the maker space to be an object of study in the Materials and Design Laboratory in the search for more sustainable actions, either by the practice of waste recycling more mainly through the identification of new meanings that can expand the user experience with this material. This study has as objective to experiment mechanical recycling techniques using PLA and ABS polymers obtained from 3D printers, and propose a new approach for this material, based on MDD methodology.

To contextualize the subject, Sect. 6.1.1 addresses the context of Do It Yourself (DIY) movement like popularity aspect of 3D printing, then Sect. 6.1.2 introduce the basic concepts and definitions about the additive manufacturing and its main techniques and materials used, with emphasis on Fusion Deposition Modeling technique. Section 6.1.3 brings the scenario of the performance of a maker space at university during the COVID-19 pandemic, in the production of face shield for the public health service. Section 6.1.4 deals with the generation of waste in the 3D printing process and the recycling process as a sustainable alternative to reduce the impact of waste.

Section 6.1.5 talks about the use of lignocellulosic fibers as reinforcement in PLA recycled matrix, point the vegetal fibers as a potential reinforce to recycled matrix from PLA. Section 6.1.6 brings the concepts of methodological proposal based on Material-Driven Design and how does this approach work. The methodology chapter presents the recycling proceedings and composites samples preparation. The results chapter presents the recycled samples of PLA and ABS, PLA/lignocellulosic fibers composites, and the characterization material.

### *6.1.1   The Maker Culture*

The maker culture that emerges because of the industrialization context, which reduced man's control to the manufacture of his own artifacts. For many years, the control of the production process and generation of products was concentrated in the largest of large industries, and this movement aims to break with this reality [9]. One of the aspects of this culture is the DIY which includes the modification, use and adaptation of existing materials for the manufacture of something through a series of creative activities, and its techniques can be shared and coded for reproduction. and improvement by others [10]. Historians indicate that the beginning of this movement occurred with the artistic movement Arts and Crafts, which appeared in the second half of the nineteenth century, in response to the industrialization and consequent devaluation of manual works. It brought a philosophical basis to manual crafts, which began to be valued again with the resumption of traditional methods, respect for material and craftsmanship [11]. The maker movement has three relevant impact characteristics, such as the use of digital desktop tools for the development and prototyping of new product projects, the culture of project sharing and collaboration in online communities, in addition to the adoption of common file formats design, allowing anyone to submit their designs for professional manufacturing services [12]. These questions serve as a basis for the manufacture of additives, and the issues of the maker movement are directly linked to the manufacturing and development process of this work as well.

## 6.1.2 Additive Manufacturing (AM): Technologies and Materials

Additive manufacturing (AM) also known as rapid prototyping is a technique to fabricate prototypes and things. A prototype is a physical model, a three-dimensional representation that allows us to explore, understand, communicate and evaluate the product alternatives generated in the design process, resulting in more consistent data for project team decision making [13, 14]. The models are used in several stages of the development of new products and can be an excellent way to introduce a new product to the market. During the design process, the models help the designer to develop new ideas mainly when it comes to products with three-dimensional complexity [15]. The intensification of production and competition between industries promoted the development of technologies that enable the reduction of time and costs in the development of projects [13] such as additive manufacturing, which is a recent and innovative technology that has been in use since the 1990s to help in the product development process [16]. The rapid prototype can work with parts of various shapes, complex geometries that would hardly be possible to obtain conventionally with low investment and reduced time. Currently, additive manufacturing is used not only to make prototypes but also to manufacture a wide variety of parts usually on a low scale of production or customized, such as home appliance parts, automotive parts, lamps, toys, jewels, and others [17, 18]. The additive manufacture has gained more prominence with the emergence of simpler and more accessible machines known as 3D printers [17].

The popularity of additive manufacturing technologies has generated a demand for new materials that could reduce costs and improve machine performance and prototype quality. The possibility of building complex shapes with less time and costs, and easy access to machines featured what was called the 4° industrial revolution and allowed that the additive manufacturing technologies were widely used in the health field, construction, sports equipment, art, and design, configuring an interdisciplinary communication scenario, sharing ideas and stimulating creativity [13, 18–20]. Next, we will highlight the main additive manufacturing technologies its main features.

### 6.1.2.1 Stereolithography (SLA)

Stereolithography was developed in 1986 and is considered the precursor process of additive manufacturing, which works with a platform that moves vertically while the object is solidified. The process starts with a liquid photopolymer that is placed in a vat and is selectively cured by a light-activated [2]. After the end of the process, the supports are removed and the object is finished and prepared for the curing stage of the resin that completes the polymerization of the model [14, 17]. The models made by stereolithography are suitable for parts with low mechanical stress or art parts.

### 6.1.2.2   Selective Laser Sintering (SLS)

A laser beam is used to sinter and solidify the pulverized material spread on a movable table. The materials most used are polymers or metals, and there are efforts to develop new materials to meet industry demand [2, 14, 17]. This technique has a promising future in the automotive industry for the rapid manufacturing of metal components.

### 6.1.2.3   Binder Jetting (BJ)

Binder Jetting is a technique that works such as a conventional print using an extrusion nozzle that delivers a binding agent on a layer of powder material spread over the construction table. The materials are usually a ceramic, plaster, or polymer, and the binding agent is a photosensitive resin that is cured instantly [2, 14, 17, 21]. The biocompatible material can be used to meet healthcare demands like prostheses [17].

### 6.1.2.4   Fusion Deposition Modeling (FDM)

Fusion Deposition Modeling is an additive manufacturing technology that builds parts by heating and extruding thermoplastic polymer filaments through a small nozzle on a machine that is known as a 3D printer [22]. The available materials for this machine are limited to thermoplastic polymers such as PLA, ABS, poly (ethylene terephthalate glycol) (PETG), polyamide (PA), polypropylene (PP), Nylon, and polycarbonate (PC) [2, 21, 23]. The FDM is the most popular among additive manufacturing technologies, because of the low cost of 3D printers, ease of operation, and availability of raw material. The thermoplastic filaments used in 3D printers have usually 1.75 mm diameter. The most used polymeric filaments for the fusion deposition modeling technique are Poly (lactic acid) (PLA) and Acrylonitrile Butadiene Styrene (ABS) [23, 24]. What differentiates additive manufacturing technologies available to industry is mainly the physical aspects of the material used (liquid resin, powder, filament), the costs, material availability, and finishing possibility [25]. Table 6.1 presented the AM technologies and some of its features.

Below are presented the two most used filaments in 3D printers and their main characteristics.

**Table 6.1** Additive manufacture technologies and materials used

| Technique | Material-based | Applications | Reference |
| --- | --- | --- | --- |
| Stereolitography (SLA) | Liquidphotopolymers | Biomedical Prototyping | [2] |
| Selective laser sintering (SLS) | Powder; Alluminium alloy, steel, titanium alloy | Automotive parts Aerospace and medical industries | [26] |
| Binder jetting (BJ) | Powder Aluminum oxide, ceramic, binder materials | Biomedical Estructures Buildings | [21, 27] |
| Fusion deposition modeling (FDM) | Thermoplastic polymers PLA, ABS, PETG, PC, PP, PA | Design, art, prostheses for healthcare, toys | [21, 28] |

### 6.1.2.5 Poly (Lactic Acid) (PLA)

Poly (lactic acid) is a biodegradable polymer that comes from renewable resources, being produced from the fermentation of agricultural resources, such as corn, is recyclable and compostable, and has applications in packaging. Its initial use was in biomedical applications due to its biodegradability, biocompatibility, and non-toxicity [29–31]. PLA has the broadest range of applications and due to its versatility, it meets several demands of the industry, being able to turn into transparent films, fibers, or injection molded [32], and currently, it has been very popular as filaments in 3D printing technologies [33]. As a biomaterial the polylactic acid (PLA) finds applications in sutures, bone fixation, drug delivery microsphere, and others [31], and, in spite of this interesting features for the industry their production costs are still high [32]. Moreover, Printed PLA preferred due to high tensile strength and less shrinkage with proper selection of process parameters of FDM Machine [34, 35].

### 6.1.2.6 Acrylonitrile Butadiene Styrene (ABS)

The Acrylonitrile Butadiene Styrene is a polymer from fossil sources and is the second most used filament in a 3D Printer, needing a heated table during the impression process. Its mechanical properties are superior compared to PLA, but it is not biodegradable [36]. During the impression process, the ABS can generate toxic smoke, beside can undergo deformation during and affect the impression due to its sensitivity to temperature variation, being preferable that the process occurs in closed printers. It also can absorb water from humidity in the air that can cause prints to varying slightly [37].

ABS has applications in fields of industry, such as home appliances, automotive parts, toys, and filaments for 3D impression [38]. The ABS filament is one of the most used polymers in 3D printing due to its mechanical properties, thermal resistance, lightweight, and to allow a good finish on the pieces [38, 39]. Table 6.2 shows the main characteristics of the PLA and ABS polymers used in additive manufacture.

**Table 6.2**  Characteristics of PLA and ABS polymers

| Polymer | Chemical structure | Melting temperature (°C) | Density (g/cm³) | Reference |
|---------|--------------------|--------------------------|-----------------|-----------|
| PLA | | 170–180 | 1.21 | [23, 33] |
| ABS | | 175 | 1.01–1.09 | [40] |

### 6.1.3   A Maker Space in an Academic Environment: Technology Incubator Network of the University of State of Pará

The Technology Incubator Network of the University of State of Pará (RITU) is a technology incubator that houses projects aimed at the technology sector, design, services, among others. It has an important role in teaching entrepreneurship in the academic community of the Center for Natural Sciences and Technology of the university. RITU also promote activities that are linked with creativity, collaborative problem solving, product development, prototyping, and entrepreneurship, in an interdisciplinary scenario. With its annual coworking notice, RITU selects projects of students of undergraduate courses such as Production Engineering, Environmental Engineering, Forest Engineering, Design, and Food Technology. The selected projects can receive, for six months, advice, and training to develop their enterprise project. In your Laboratory Concept in Engineering, Processes, and Technology (ConcEPT), the rapid prototyping sector works with 3D printers' machines that use PLA and ABS polymer. In the COVID-19 pandemic, this laboratory has been used to manufacture face shields for the Public Health System (SUS). With the intensive use of 3D printers, there was an increase in the generation of waste and the need for further studies in order to reduce these residues.

### 6.1.4   Waste Recycling as a Sustainability Alternative for 3D Printing

The concern with environmental aspects has grown over the past 20 years and originated specific environmental legislation for some industry sectors. Nowadays materials and production processes are analyzed according to their impact on the environment, which led to the valorization of biodegradable or recyclable materials intending to reduce the demand for landfills. Sustainability concepts have become increasingly relevant in many industry sectors. Recently, additive manufacturing technologies have also started to be studied in terms of its environmental aspects and the importance of sustainability in the design process [2, 36, 41–45]. Specifically, in the FDM technique, a few works were found mainly about thermoplastic waste recycling [45], another proposing the use of recycled raw material [24, 46]. The waste in 3D printer machines can be generated in the following cases [47]:

- In the parameter setting phase (table temperature adjustment, extruder nozzle temperature adjustment, extruder nozzle displacement);
- During the coil change or extruder nozzle cleaning;
- When there is a power supply failure that interrupts the printing process;
- During the removal of support (post-processing activities).

In most maker spaces recycling waste is still not given due importance. The waste from ConcEPT/RITU laboratory is thermoplastic polymers (PLA, ABS) and is collected and prepared for the recycling process by the Materials and Design Laboratory of the University of State of Pará. This process is considered secondary recycling because the waste must be collected, separated, crushed, washed, and dried for mechanical recycling procedures [48]. In this scenario, the recycling process is motivated by teaching and research activities developed in the Materials and Design Laboratory, where are conducted mechanical recycling experiments, transforming waste into an object of study for design students that has the opportunity to exercise sustainable design concepts [46, 48].

## 6.1.5 Thermoplastic Composites Reinforced with Lignocellulosic Fibers

Polymeric composites reinforced with fibers are materials of interest to the industry due to the capacity of ability to design properties and the possibility to use lignocellulosic fibers, adding environmentally friendly characteristics to the material/product. Vegetal fiber-reinforced thermoplastic composites have advantages, such as low cost, low density, less abrasiveness, and have found application in furniture, packaging, and automobile industries [49]. PLA is a biopolymer that can work as a thermoplastic matrix reinforced with any natural fiber, some researchers have been investigated about this topic, that used cellulose nanofibers from kenaf to reinforce a PLA matrix prepared banana fibers to reinforced PLA membranes, used four types of PLA to conduct experiments with wood fiber/PLA composites, prepared a matrix with PLA and pine resin does develop an appropriate blend to made composites reinforced with açaí fibers [8, 14, 47, 50]. A previous work conducted at the Materials and Design Laboratory of the university, PLA waste were collected from ConcEPT/RITU laboratory and recycling showing the mechanical recycling possibilities of this thermoplastic [46].

## 6.1.6 Material-Driven Design (MDD)

With technology, the advent and creation of new materials, several changes in the formal aspects and the use of products have occurred, so that the products are not only perceived by the functional and aesthetic issue, but also analyzed taking into account emotional and sentimental factors [51]. The Material-Driven Design methodology (MDD) is a methodology aimed at studying materials based on experience, that is, the material developed in addition to the functional issue, your project also aims to raise experiences, predefined in the material study process [6]. The material development process is geared beyond technical aspects since the user experience is an extremely extravagant factor in the use of a product or service, the materials develop in the user various associations of ideas and concepts, or even sensations [52]. Another fundamental factor for the study of materials in products is the environmental issue, since the selection of the material influences the sustainability of the product, because of the resulting environmental problems and the pressure for new sustainable alternatives [53]. So that in the selection of materials these steps are analyzed to define the choice of a sustainable material not only by its origin but also by its processing. Thus, the material contributes to the generation of alternatives in the design process and thus in the execution of the requirements in the project [52], based on the implementation of a methodology centered on the material. That said, in the Design methodology conducted by the material, the design process is centered on the study of a material, being possible through the study of its properties, limitations and opportunities [6]. The MDD methodology requires understanding what the material offers in terms of its function, utility and experience [54] for this it is necessary to understand in which scenario this material is inserted, as well as its behavior in different scenarios and how it is perceived by a group of people [6]. The MDD methodology can be applied within three possible scenarios: for a material known with a developed sample that already contains derived meanings, for an unknown material with a developed sample, but without materialized meanings or for a new material proposal with a semi-developed sample or still in an exploratory phase [6]. After understanding the scenario to which the material studied belongs, it is the four stages that make up the methodology, which will be built through interaction and reflections as to what was experienced and experienced with the material. The first stage focuses on understanding and developing repertoire about this, called technical and experiential characterization of the material. To obtain the technical characterization, the researcher's interaction with the material is necessary and indispensable, being able to cut it, knead it, burn it, as well as other manipulations that contribute to the understanding of the technical properties and manufacturing processes of the product same. Experiential characterization, in turn, involves the four levels of experiences, initially experienced by the researcher and later presented to users, always based on interaction and experience with the material. Still in this phase, it is recommended to carry out a benchmarking of materials, aiming to fit it within a group of similar materials, visualizing possible areas of applications, as well as reflecting on the technical specifications and experience qualities perceived in similar ones and their applications [6]. At this moment, there is an understanding of what is particular to the material studied, as well as what exists beyond it and how people relate to these other possibilities. After the first stage has been completed, and based on the results obtained, it is possible to proceed to the next stage or skip to step 3 or 4, depending on the result to be obtained. From the discoveries obtained during the characterization of the material, sufficient understanding is acquired to create the vision of the materials experience, pertinent to the second stage, which is developed through the reflection on the singularities of the material, its possible contributions and scenarios positively impacted after its implementation. As well as, building hypotheses as to how the user's experience with the material would occur, from a sensory, emotional, interpretive and performative point of view, based on the findings of the previous phase, to obtain a basis for the construction of one or more vision statements and a possible user-product interaction [6]. From this, a possible future scenario is constructed and visualized, in which it is possible to understand the action of the material within a social context, of use, interaction and, consequently, of experience. To complement this, in the third stage—manifesting material experience patterns—in a brainstorm session, of the vision and interaction, the meanings corresponding to how the material is supposed to be interpreted by the user are identified (interpretive level of experience), with the assistance from the study with users, it will be possible to compare and analyze what was foreseen in the vision and what the material and its formal qualities raise in the public during the experimentation [6]. With this, there is the necessary support to understand the existing interrelationships, the patterns that manifest themselves during the interaction and what is the propitious scenario for them to occur. The final step, in turn, is aimed at conceptualizing the material and product, explains that, having or not an idea of a product to be developed, this stage is intended to work with the material itself, manipulating it, to obtain satisfactory results, in line with what was envisioned in the Vision and Standards of Material Experiences, respectively, stages 2 and 3. It also points out that the conceptualization of the material occurs through the creation, analysis and selection of alternatives and idealization of the final concept, for later prototyping and testing. Therefore, having the material and its concept properly developed and within what was intended, it is possible to start the process of conceptualizing the product and, consequently, the construction of it. From what has been seen, such methodology has been widely used in the generation of new products and materials, uses coffee residues to generate not only the material but also the concept for the product, as well as using Mycelium as a raw material in the development of the material [6, 54]. Another work carried out using the Material-Driven Design methodology is the recycling of plastic and generation of concepts and product alternatives for the material, resulting in recycled plastic lamps [53] In this way, it is possible to observe several materials and products developed from this methodology, thus, the present work uses the MDD methodology in the generation of alternatives for 3D printing residues.

## 6.2   Methodology

### 6.2.1   Mechanicalrecyclingprocess

PLA and ABS wastes, Fig. 6.1, were collected by the ConcEPT/RITU laboratory to be recycled at the Materials and Design Laboratory. The color of PLA is orange, the color of ABS is blue. For the mechanical recycling process, the polymers waste were cut into small pieces, washed, and dried, then were submitted to the melting process until they reach the melting temperature, 170 °C for the PLA, and 175 °C for the ABS.

Two equipments were used to foundry the waste polymer: In the first one, an oven was used and the waste was placed in a silicone mold to reach the melting temperature, the mold was removed from the oven and the samples were naturally cooled. In the second one, a pan was used to melt the polymer waste and then poured into silicone molds.

### 6.2.2   Preparation of Composite Samples

To verify the performance of recycled PLA to form composite plate's three natural fibers were selected to reinforce the PLA matrix:

**Fig. 6.1** PLA (orange) and ABS (blue) wastes. *Source* Authors, 2020

- Untreated Jute fiber (Corchorus capsularis) provided by CastanhalTêxtil Company and cut with 1.5 cm length;
- Untreated Açaí fiber (Euterpe Oleracea M.) provided by Materials and Design Laboratory;
- Untreated wood flour provided by Product Design Laboratory.

The fibers were mixed with PLA matrix using a pan and poured into the mold until cooling. Were consolidated three composite plates with PLA as matrix: wood flour (PLA/wood flour), with jute fiber (PLA/jute fiber), and with açaí fiber (PLA/açaí fiber). All the samples were organized to Material-Driven Design analyze.

## 6.3   Results

Based on what has already been discussed about the MDD methodology, the recycling of 3D printing waste was used, among which are PLA and ABS, which according to the methodology fall into the first scenario because they are relatively known materials, since they are more polymers used in additive manufacturing due to their properties, as well as, the research seeks to find new meanings and applications to that beyond 3D printing. Therefore, in the first stage of the methodology, the material was studied both in relation to its mechanical and technical properties and in relation to the manufacturing processes carried out to obtain the samples.

The main properties, qualities and restrictions of the materials were studied, which are already available in the literature. ABS is a copolymer derived from the reaction of acrylonitrile, butadiene and styrene monomers, and in relation to its mechanical properties and considered light, durable, slightly flexible, in addition to withstanding high temperatures; however, it presents as a disadvantage the generation of toxic smoke during printing process and its origin in oil and therefore a non-biodegradable material [36] Poly (lactic acid) (PLA), on the other hand, is a biodegradable material, synthesized from lactic acid from renewable sources, as it is also considered a non-toxic and easy to work material; however, it presents the disadvantage of being a brittle material [36]. As for the process of manufacturing the samples, the recycling of waste occurred in two ways, the plastic was heated in an artisanal manner in both an oven and a stove. Thus, the recycling of waste begins with the selection of waste to be recycled, in which the separation of ABS and PLA waste was carried out, which were separated according to color, with the blue ones being ABS and the orange PLA. Subsequently, the separation proceeded to the cleaning and shredding stage, in which the residues are cut into small pieces to advance to the heating stage of the material, that is, the smelting of the waste. The heating of the residues occurred in two ways, in the first, the residues already cut were deposited in a silicone mold and heated in an oven and pressed after the removal of the silicone mold from the oven. In the second form of heating, the residues were subjected to constant heating and stirring in a pan and deposited in a silicone mold and then pressed (Fig. 6.2).

**Fig. 6.2** PLA recycling process. *Source* Authors, 2020

Among the ways of recycling the residues, the most convenient process varied according to the material, the PLA residues obtained greater melting when heated in the oven, their manufacture in the pan took place more laboriously taking into account the need for constant agitation of the material, but also, when recycled in the pot resulted in the loss of most of the separated residues, since when heated in the pot, a large amount of the residues present has degraded, leaving a small portion of the material to be pressed. Other different aspects regarding the samples developed for each recycling mode, were color and shape, while the PLA had its orange color preserved when heated in the oven (Fig. 6.3), it acquired brown color (Fig. 6.5) being heated in the pan. As well as in relation to the shape, when recycled in the oven, the PLA acquired the shape of the mold, square, while it did not obtain an exact shape in the other mode even though it was deposited in a mold later.

For ABS, the most favorable process for manufacturing the samples was recycling in the pan. The ABS residues obtained greater melting when subjected to heating and agitation in the pan, for their manufacture in the oven, the ABS did not reach its melting point, taking longer to heat in the oven, as it did not obtain complete melting of all the residues disposed in the silicone mold. The color and shape, however, did not diversify much from one process to another, in the two recycling processes carried out, the blue color remained even after heating the residues, as well as when heated in the oven it obtained a square shape, which it acquired in the shape of the mold, however, showed flaws and imperfections on its surface, since it did not melt completely. When cast in the pan, the ABS acquired a square shape of the mold which was deposited after heating (Fig. 6.4).

In another experiment, PLA was recycled in the pot, and its varied core waste resulted in brown-colored knots due to material degradation when recycled in the

**Fig. 6.3** Samples of recycled
PLA. *Source* Authors, 2020

**Fig. 6.4** Samples of recycled
ABS. *Source* Authors, 2020

pot at high temperature. As also happens with the manufacture of the PLA sample
recycled with Jute fiber, in which the orange PLA residues when recycled in the pan
obtained a yellow color (Fig. 6.5), with fibers being added later to the casting of the
material, resulting in a yellow PLA sample with jute fiber.

**Fig. 6.5** Samples of recycled PLA with Jute fibers. *Source* Authors, 2020

The PLA samples were made with other materials besides fibers (Fig. 6.6), such as wood powder, their sample acquired a lighter brown color than the other samples that obtained this same color, and its surface presents a rough texture.

**Fig. 6.6** Sample of PLA with wood powder. *Source* Authors, 2020

**Fig. 6.7** PLA (brown) and
ABS (blue) samples. *Source*
Authors, 2020

The PLA and Açaí fiber recycling process took place based on the previous procedures and two tests were carried out. First, there was a mixture of the fiber and the material during the melting point of the PLA. Subsequently, the fibers were dispersed in the mold and the polymeric matrix was applied to the mold. In both materials, the use of Açaí fiber as reinforcement was very evident. When mixing the PLA, the material presented roughness and care was taken so that the fibers did not suffer degradation due to the presence of heat. In the mixture with fiber in the mold, the plant material stood out as predominant characteristics of its natural shape. The final appearance of the samples varied according to the material used, as well as the manufacturing process to which it was submitted. Orange PLA samples range from a smooth surface, due to the pressing process, and glossy to a rough texture with varied deformations and opaque or on a surface with both characteristics. While some samples had their color preserved, some had changed in their color, as is the case with the orange PLA, which showed brown variations (Fig. 6.7), due to the recycling process carried out in the stove, with deformations of crystallized appearance on one face and with the another completely uniform. Samples that had no color changes were heated in the oven. The color change also occurred with a blue ABS (Fig. 6.7) sample which showed a mixture of blue and black, very similar to brush strokes, shiny, together with an oily texture, showing cracks along with its surface.

Thus, it can be said that the changes made during the manufacturing process of each sample were of vital importance in the tactile and visual variations presented. During the work and handling of materials with vegetable fibers, the tactical and visual issue is very important when you want to reach an audience with demands for products or services aligned with sustainability, considering the natural factor of your raw material.

## 6.4   Experimental Characterization of the Material

Due to the COVID-19 pandemic, face-to-face meetings for the study of users could not occur with a large sample of people, this stage is restricted to interviews with small groups, groups of 4 people in person, and the application online question-naires with a group of 23 people. As a result, variations can be seen in the way the samples are perceived, as some groups have physically interacted with the material, while others have only evaluated it visually. Of the opinions obtained around the samples, texture and color got the most attention from the public, however, color was the most pleasing aspect to the public eye, while a small part indicated the texture as an unpleasant aspect and most people indicated that does not consider any aspect unpleasant. Regarding the emotions aroused, curiosity was the emotion that most people pointed out about the samples, followed immediately by surprise and disgust. In terms of interaction, most of the public indicated that they would like to touch and squeeze the sample if they had contact with it, the table below details the responses indicated in the questionnaire regarding the four experiential levels. It is important to note that due to the varied universe of samples developed, with different sensory characteristics, to the public's perceptions (Fig. 6.8).

## 6.5   Material Benchmarking

For the second stage, it is still recommended to carry out a benchmarking of materials, to position it within a group of similar, comprising possible areas of application, observing the experiential issues raised by these, and also under-standing the main changes and strategies that come consolidating in the design field [54]. A structure similar to that presented in the article The Plastic Bakery: A case of Material-Driven Design to demonstrate the data in Table 6.3.

During the benchmarking, it was noticed several applications in products that arise from the need to give a new direction to plastic waste, reframing materials that before would end up in the garbage and that now collaborate in the generation of jobs and income for members of cooperatives. Selective collection, as well as the continuity of environmental preservation projects. The products from the plastic waste make evident the sustainable purpose and to stimulate the community spirit of environmental preservation, it is clear that the public consuming these artifacts does not buy the product itself, they buy the purpose behind it, the history of the people involved in its development and how its existence has a positive impact on the environment. There is still little to be seen, from a commercial point of view, products derived from additive manufacturing residues, however, there are already studies on their potential. Thus, the users' responses in the questionnaire led to the selection of two samples for the following stage, PLA sample with açaí fibers and blue ABS, which drew the attention of the interviewees due to their appearance,

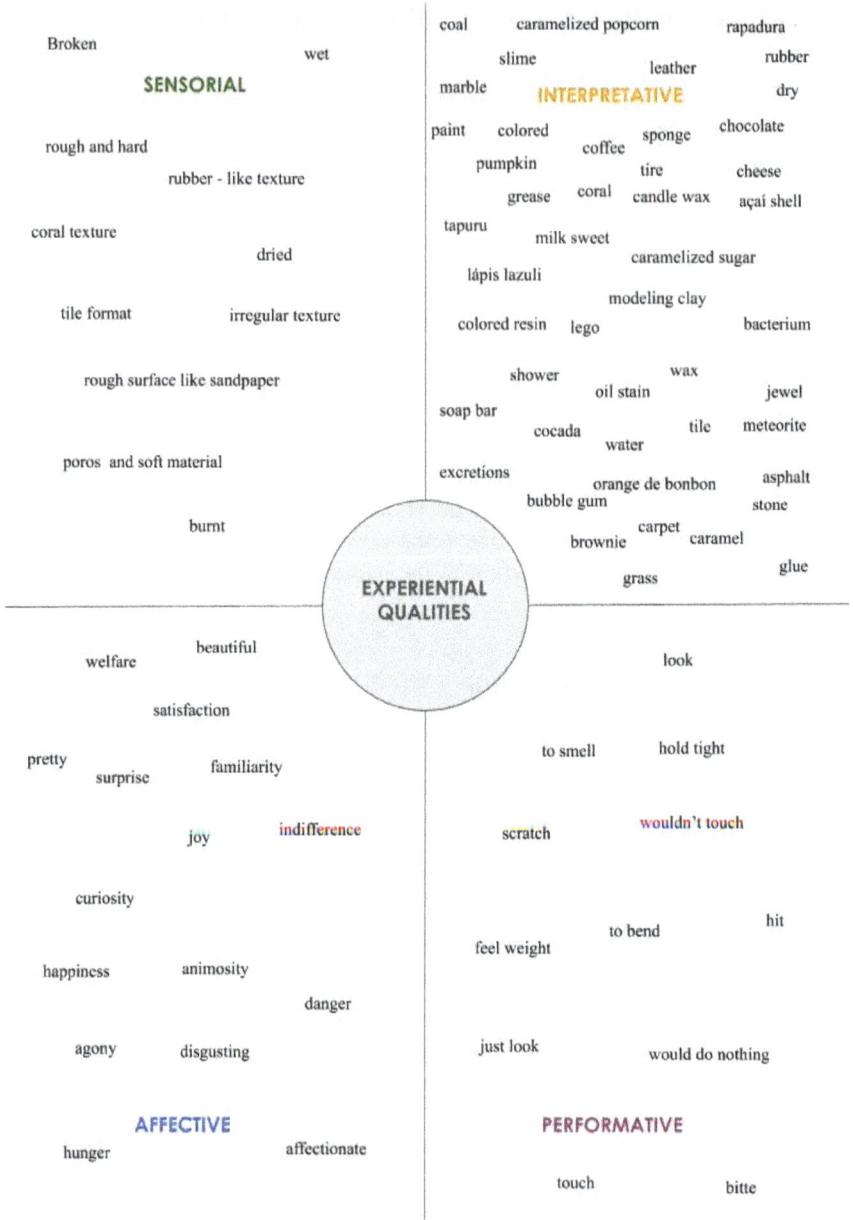

**Fig. 6.8** Mind map with users' perception. *Source* Authors, 2020

texture, shape, and the meanings they refer, so that in the next stage, possible suggestions are presented regarding the use of the material, based on people's perception of it.

**Table 6.3** Benchmarking of materials, characteristics of PLA and ABS polymers

| Application Picture/ Sample |  Source: Plasacre [55] |  Source: Spark&Burnish [56] |  Source: SmilePlastic [57] |
|---|---|---|---|
| Name | Plasacre tile [55] | Door Handle—Ocean Platic Knob [56] | Handmade Waste Panel [57] |
| Manufacturer | Plasacre | Spark and Burnish | SmilePlastic |
| Composition | Polypropylene, additive with Anti-UV, pigmented, produced by the injection process through Chinese injectors of the HAITIAN brand | Plastic waste taken from the ocean | Plastic waste |
| Experiential qualities | Bright, uniform, smooth surface, varied colors | Imitates marble (sophisticated) or concrete, opaque aspect, they seem to be heavy | Interesting, fun chromatic variations, unique visual compositions, lackluster, rigid and resistant |
| Applications | Roofs | Door handles or wall hooks | Panels, table and bench tops, doors, cabinets, facades, etc. |
| Activities | Coverage of residential spaces | Decorative/utility item for residential spaces | Decorative items from shops, exhibitions and offices |
| Ultimate purpose | Give a new direction to the plastic waste that would go to landfills, transforming them into new products, contribute to the existence of the Solid Waste Treatment Unit, the selective collection and cooperatives of waste collectors involved, generating jobs and income in the region | Offer ecological options, contribute to two Australian non-profit organizations for selective collection and protection of the oceans | Modify the view of waste, demonstrate the potential of recycling and the products that may come from it. Inspire about sustainability and recycling |

*Source* Author, 2020

## 6.6 Creating Material Concepts

Two samples were selected for product development, the blue ABS sample, and the PLA sample with açaí fibers (Fig. 6.9), presented before the benchmarking. The sample of PLA with added fibers in its composition aroused both curiosity and negative reactions such as disgust and fear due to its fibers exposed externally; however, it stood out as to other samples due to its difference in texture and color.

The blue ABS sample (Fig. 6.10), on the other hand, was as prominent as the PLA sample, with its color variation being pointed out by the interviewees, which gave it an interesting appearance of a mixture of colors, being compared with brush strokes, galaxies, lava and oil stains, this sample having a greater positive reaction with the public, awakening from reactions of surprise and curiosity to satisfaction.

Thus, as a suggestion for the sample of PLA with added fibers is its possible application in the packaging sector, the use of fibers in packaging being a proposal that already exists in the market, however, the use of PLA with fibers, a composite material, can benefit packaging applications. For the blue sample, according to the perception of most of the interviewees in the questionnaire, regarding the associations related to the blue ABS sample, this was associated with jewelry, gems and even compared to Lapis lazuli, from there the suggestion of application for that, is its use in jewelry and accessories.

In this way, it is possible to observe the different appearances, in what consists of shape, color and texture that the material reaches in its recycling process, either due to the difference in manufacturing processes or the addition of other materials to the waste, as is the case with fibers. As well as it is also possible to perceive the different perceptions of people with samples originating from 3D printing residues, which are perceived in different ways than a plastic used in additive manufacturing. And through the different understandings about the material, as well as the different

**Fig. 6.9** PLA samples with açaí fibers selected for the work. *Source* Authors, 2020

**Fig. 6.10** Blue ABS samples selected for the work. *Source* Authors, 2020

performances it presents, it is possible to develop different alternatives and application proposals regarding the use of the 3D printed material for rapid tool for machining as explored by previous work [58].

## 6.7 Conclusion

There is no doubt that AM and its several techniques revolutionized in the field of consumer goods development, especially when it comes to reducing waste generation when compared to traditional techniques like machining, and the reduced time to launch the new product, besides expanding communication in the design process. Among the AM technologies, the Fusion Deposition Modeling is the most popular technique due to the low cost of 3D printers which used thermoplastic polymer filaments such as PLA and ABS. In this work, the partnership between ConcEPT/ RITU and Materials and Design Laboratory brought the capacity for interaction and interdisciplinarity as important factors in the search for sustainability.

The selective collection implemented in maker spaces and the laboratories of teaching and research institutions can contribute to the recycling of waste from 3D printers and make these spaces more sustainable. PLA and ABS recycling proved to be viable, but care must be taken when recycling ABS waste due to the toxic smoke that the polymer produces during processing. The use of recycled waste as a polymeric matrix in the development of composites reinforced with vegetable fibers expands the sustainability aspects of the material. The MDD methodology provides it to be adequate to identify and understand the possibilities of the material and contribute to the design process. Besides, it highlights the alignment with the ideas of the maker movement, in which the productive process of recycling and material can be more accessible. These possibilities must be explored by design for the application of the material in products that meet contemporary demands. The user's interest in samples of recycled PLA with Açaí fiber demonstrates the possibility of applying this material in products, in addition to the opportunity to use vegetable fibers from residues to form other types of materials. In this case, evaluating the possible mechanical, thermal advantages, and the possibility of biodegradation of this new material. For this, it is necessary to also make the technical characterization of the material, in addition to the experimental characterization proposed by the MDD methodology. Also, complementing the methodological steps of the MDD, the possibility of applying the perception test to users in applied products should be considered. Assess whether the designer, or the design team, met the material's expectations and whether its application in products would be accepted or rejected.

# References

1. Villamil, C., Nylander, J., Hallstedt, S.I., Schulte, J., Watz, M.: Additive manufacturing from a strategic sustainability perspective. Int. Des. Conf. Des. Methods 1381–1392 (2018)
2. Kellens, K., Baumers, M., Gutowski, T.G., Flanagan, W., Lifset, R., Duflou, J.R.: Environmental dimensions of additive manufacturing. Mapping applications domains and their environmental implications. J. Ind. Ecol. (2017)
3. Santana, L., Alves, J.L., Sabino Netto, A.C., Merlini, C.: Estudo comparativo entre PETG e PLA para Impressão 3D através de caracterização térmica, química e mecânica. RevistaMatéria, v. 23, n. 4 (2018)
4. D'Ambrières, W.: Plastics recycling worldwide: current overview and desirable changes. J. Field Actions Special, Issue **19**, 11–21 (2019)
5. Majumdar, P., Karana, E., Ghazal, S., Sonneveld, M.H.: The plastic bakery: a case of material driven design. EKSIG 2017: Alive. Active. Adaptive 1–13 (2017)
6. Karana, E., Barati, B., Rognoli, V., Van Der Lan, A.Z.: Material driven design (MDD): a method to Design for material experiences. Int. J. Des. **9**(2), 35–54 (2015)
7. Leite, C.D.P., Teixeira, L.F., Cohen, L.A.F., Santos, N.S.S.: Recovery and recycling of a biopolymer as an alternative of sustainability for 3D printing. In: Proceedings of the 3 rdLeNS World Distributed Conference (on line). Milano, Mexico City, Bangalore, Curitiba, Cape Town, vol. 1 (2019)
8. Santos, N., Silva, M.R., Lino, J.: Reinforcement of a biopolymer matrix bylignocellulosic agro-waste. Procedia Eng. (2017)

9. Illich, I.: Tools ForConviviatility. Marion Boyars, Londres (1973)
10. Buechley, A., et al.: DIY for CHI: methods, communities, and values of reuse and customization. In: CHI'09: 2009: New York (2009)
11. Koplos, J., Metcalf, B.: Makers: A History of American Studio Craft. The University of North Carolina Press, Hendersonville (2010)
12. Hatch, M.: The Maker Movemet Manifesto: Rules for Innovation in the New World of Crafters, Hackers, and Tinkerers. MsGraw Hill Education (2014)
13. Alcoforado, M.G., Paschoarelli, L.C., Silva, J.C.P.: Metodologia centrada nos protótipos: um caminho para inclusão de usuários no processo de design. In: 15º Ergodesign, USIHC (2015)
14. Pradella, M.P., Folle, L.F.: Análise de mercado sobre tecnologias de prototipagem rápida por adição de material. In: anais do 11º P&D Design, Congresso Brasileiro de Pesquisa e Desenvolvimento em Design, Gramado (2014)
15. Ferroli, P.C.M.: Uso de modelos e protótipos para auxílio na análise da sustentabilidade no Design de produtos. GEPROS. Gestão da Produção, Operações eSistemas, Ano 7, nº 3, (2012)
16. Bruscato, U.M., Brendler, C.F., Viaro, F.S., Teixeira, F.G., Silva, R.P.: Uso Da Fabricação Digital E Prototipagem No Desenvolvimento Do Projeto Do Produto: Análises Do Produto Através De Simulações Digitais. SiGraDi (2013)
17. Mancanares, C.G.: Proposta De Um Método De Seleção Do Processo De Prototipagem Rápida Para Fabricar Uma Peça A Partir De Especificações Técnicas Enegep (2013)
18. Chicca Jr., N.A., Castillo, L. G.: Impressão 3D na cultura do design contemporâneo. In: Anais do 11º. P&D
19. Rodrigues JR, J.L., Cruz, L.M.S., Sarmanho, A.P.S.: Impressora 3d No Desenvolvimento De Pesquisas Com Próteses. Rev. Interinst. Bras. Ter. Ocup. Rio De Janeiro (2018)
20. Lussenburg, K., Velden, N.V.D., Doubrovski, Z., Geraedts, J., Karana, E.: Designing with 3Dprinted Textiles. A case study of Material Driven Design (2014)
21. Sandeep, C.D.: Comparison and analysis of different 3D printingtechniques. Int. J. Latest Trends Eng. Technol.
22. Alafaghani, A., Qattawi, A., Ablat, M.A.: Design consideration for additive manufacturing: fused deposition modelling. Open J. Appl. Sci. (2017)
23. Santana, L., Alves, J.L., Sabino Netto, A.C., Merlini, C.: Estudo comparativo entre PETG e PLA para Impressão 3D através de caracterização térmica, química e mecânica. Revista Matéria (2018)
24. Paiva, D.A., Souza, E.W., Ferreira, A.C.B., Bonse, B.C.: Compostos de ABS virgem e reciclado para impressão em 3D. In: Anais do 13º Congresso brasileiro de Polímeros (2015)
25. Santos, L.M., Rocha, D.S.G.M., Carneiro, M.L., Luz, M.P.: Tipos de polímeros utilizados como matéria-prima no método de manufatura aditiva por FDM: uma abordagem conceitual. XXXVIII Encontro Nacional de Engenharia de Produção (2018)
26. Villamil, C., Nylander J., Hallstedt, S.I., Schulte, J., Watz, M.: Additive manufacturing from a strategic sustainability perspective. In: International Design Conference (2018)
27. Ngo, T.D., Kashania, A., Imbalzanoa, G., Nguyena, K.T.Q., Hui, D. Additive manufacturing (3D printing): a review of materials, methods, applications and challenges. Composites Part B (2018)
28. Zuniga, J., Katsavelis, D., Peck, P., Stollberg, J., Petrykowski, M., Carson, A., Fernandez, C.: Cyborg beast: a low-cost 3D-printed prosthetic hand for children with upperlimb differences. BMC Research Notes (2015)
29. Avinc, O., Khoddami, A.: Overview of poly(lactic acid) (PLA) fibre: part i: production, properties, performance, environmental impact, and end-use applications of poly(lactic acid) fibres. Fibre Chem. (2009)
30. Pereira, R.B., Morales, A.R.: Estudo do comportamento térmico e mecânico do PLA modificado com aditivo nucleante e modificador de impacto. Polímeros (2014)
31. Lopes, M.S., Jardini, A., Macielfilho, R.: Synthesis and characterizations of poly (lactic acid) by ring-opening polymerization for biomedical applications. Chem. Eng. Trans. (2014)

32. Mohanty, A.K., Misra, M., Drzal, L.T., Selke, S.E., Harte, B.R., Hinrichsen, G.: Natural fibers, biopolymers, and biocomposites: an introduction. In: Natural Fibers, Biopolymers, and Biocomposites. Taylor & Francis Group, LLC. CRC Press (2005)
33. Santana, L., Ahrens, C.H., Sabino Neto, A.C., Oliveira, G.M.B., Merlini, C.: Avaliação da composição química e das características térmicas de filamentos de PLA para impressoras 3D de código aberto. In: anais do IX Congresso Nacional de Engenharia Mecânica, Fortaleza (CE), Brasil (2016)
34. Sandhu, K., Singh, J.P., Singh, S.: Some investigations on the tensile strength of additively manufactured polylactic acid components. In: Advances in Materials Processing, pp. 221–230. Springer, Singapore (2020)
35. Sandhu, K., Singh, S., Prakash, C.: Analysis of angular shrinkage of fused filament fabricated poly-lactic-acid prints and its relationship with other process parameters. In: IOP Conference Series: Materials Science and Engineering, Vol. 561, No. 1, p. 012058. IOP Publishing (2019)
36. Besko, M., Bilyk, C., SIEBEN, P.G.: Aspectos técnicos e nocivos dos principais filamentos usados em impressão 3D. GestãoTecnologiaInovação (2017)
37. Baker, C., Brent, D., Wilson, C., Xu, J., Thompson, L.A.: Additive manufacturing for economical, user-accessible upper-limb prosthetics. IMedPubJournals (2017)
38. Santos, L.M., Rocha, D.S.G.M., Carneiro, M.L., Luz, M.P.: Tipos de polímeros utilizados como materia prima no método de manufatura aditiva por fdm: uma abordagem conceitual. XXXVIII Encontro Nacional de Engenharia de Produção (2018)
39. Vishwakarma, S.K., Pandey, P., Gupta, N.K.: Characterization of Abs material: a review. J. Res. Mech. Eng. 7 (2017)
40. Almeida, A.M.G.: Estudo da possibilidade de utilização do polímero Acrilo, processado via impressão 3D para produção de embalagens retornáveis de fluxos logísticos para a indústria automotiva. Cefet (MG), Belo Horizonte (2016)
41. Diegel, O., Singamneni, S., Reay, S., Withell, A.: Tools for sustainable product design: additive manufacturing. J. Sustain. Dev. (2010)
42. Ferroli, P.C.M.: Uso de modelos e protótipos para auxílio na análise da sustentabilidade no Design de produtos. GEPROS (2012)
43. Angioletti, C.M., Sisca, F.G., Luglietti, R., Taisch, M., Rocca, R.: Additive manufacturing as an opportunity for supporting sustainability through implementation of circular economies. XXI Summer School "FrancescoTurco" - Industrial Systems Engineering (2016)
44. Peng, T., Kellens, K., Tanga, R., Chenc, C., Chen, G.: Sustainability of additive manufacturing: an overview on its energy demand and environmental impact. Additive Manufacturing
45. Manrique, M.R., Mendes, L.T., Laurentino, A.L., Seabra Filho, S.S. Plástico Precioso: prototipagem rápida e reciclagem de resíduos de manufatura aditiva
46. Leite, C.D.P., Teixeira, L.F., Cohen, L.A.F., Santos, N.S.S.: Recovery and recycling of a biopolymer as an alternative of sustainability for 3D printing. In: Proceedings of the 3 rdLeNS World Distributed Conference (2019)
47. Sallenave, G.C., Caldovino, G.C., Silva, F.P., Candido, L.H.A., Jacques, J.J.: Contribuições para a discussão dos resíduos gerados pelo processo de fabricação por filamento fundido (FFF). In: Design em pesquisa (2020)
48. Spinacé, A.S., De Paoli, M.A.: A Tecnologia da Reciclagem De Polímeros
49. Pereira, P.H.F., Rosa, M.F., Cioffi, O.H., Benini, K.C.C.C., Milanese, A.C., Voorwald, H.J. C., Mulinari, D.R.: Vegetal fibers in polymeric composites: a review. Polímeros (2015)
50. Oliveira, G.R., Santos, T., Pacheco, K., Grisa, A.M.C., Zeni, M.: Seletividade A Gases De Membranas de Poli (Ácido Lático) Reforçadas com Fibra de Bananeira. Revista Iberoamericana de Polímeros (2013)
51. De Moraes, D.: Metaprojeto: o design do design. Blucher, São Paulo (2010)
52. Calegari, E.P., Oliveira, B.F.: Um estudo focado na relação entre design e materiais. Projética (2013)
53. Thompson, R.: Materiais sustentáveis, processos e produção. Editora Senac, São Paulo (2015)
54. Karana, et al.: When material grows: a case study on designing (with) Mycelium-based materials. Int. J. Des. (2018)

55. Plasacre.Plasacre tile, from http://plasacre.com.br/
56. Spark & Burnish. Door Handle—Ocean Platic Knob, from https://sparkandburnish.com.au/collections/knobs/products/ocean-plastic-knob-3
57. Smile Plastic. Handmade Waste Panel, from https://smile-plastics.com/commissions/
58. Sandhu, K., Singh, G., Singh, S., Kumar, R., Prakash, C., Ramakrishna, S., Królczyk, G., Pruncu, C.I.: Surface characteristics of machined polystyrene with 3D printed thermoplastic tool. Materials **13**(12), 2729 (2020)
59. Jonoobia, M., Haruna, J., Mathew, A.P., Oksman, K.: Mechanical properties of cellulose nanofiber (CNF) reinforced polylactic acid (PLA) prepared by twin screw extrusion. Compos. Sci. Technol. (2010)
60. Majumdar, P., Karana, E., Ghazal, S., Sonneveld, M.H.: The plastic bakery: a case of material driven design. In: Proceedings of the International Conference of the DRS Special Interest Group on Experiential Knowledge and Emerging Materials (2017)

# Chapter 7
# Utilization of Agro Waste for the Fabrication of Bio Composites and Bio plastics—Towards a Sustainable Green Circular Economy

S. N. Kumar, Roopal Jain, K. Anand, and H. Ajay Kumar

**Abstract** This chapter proposes the utilization of agro-waste for the fabrication of bio composite and bio plastics. Agro waste is an efficient source for the fabrication of composite material. In the industrial, medical and agricultural sector, the natural fibres based reinforcement is gaining prominence. The natural fibres are classified based on the origin and can be categorized into plant, mineral and animal. The natural fibres have noteworthy gains over synthetic fibres. The composites and plastics based on naturally available resources are gaining importance due to the renewable and eco-friendly nature with the environment. India is blessed with a wide variety of plants and trees and the waste generated from nature when utilized properly paves a way towards sustainable development. This chapter focuses on the characteristics of some of the typical bio composites and bio plastics. The characteristics of bio composites and bio plastics depend on the treatment and process involved in the conversion of agro-waste. The applications of bio plastics and bio composites in various sectors are also highlighted in this work. The agro waste is one of the sources for the fabrication of bio composites and bio plastics and efficient utilization of agro-waste also generates rural empowerment towards a sustainable green circular economy. The agro waste based bio composites and bio plastics have significant environmental and economic benefits.

**Keywords** Bio composites · Bio plastics · Sustainable development · Rural empowerment · Natural fibres

S. N. Kumar (✉) · K. Anand
Amal Jyothi College of Engineering, Kottayam, Kerala, India

R. Jain
CSIR-NEIST, Johart, Assam, India

H. Ajay Kumar
Mar Ephraem College of Engineering and Technology, Elavuvilai, Tamilnadu, India

© The Author(s), under exclusive license to Springer Nature Switzerland AG 2022  131
K. Sandhu et al. (eds.), *Sustainability for 3D Printing*, Springer Tracts in Additive Manufacturing, https://doi.org/10.1007/978-3-030-75235-4_7

## 7.1  Introduction

A composite is a material that includes at least two consolidated constituents that are united at the macroscopic level and are not dissolvable in one another. One constituent is called reinforcing material and one in which it is embedded is known as the matrix. The fibres are widely used as the reinforcing materials and sometimes grasses are also used. The matrix material is in continuous form, for example, tapioca starch. Instances of normally discovered composites incorporate wood, where the lignin lattice is strengthened with cellulose filaments and bones in which the bone-salt plates made of calcium and phosphate particles fortify delicate collagen. The role of reinforcement in composites is to get quality, firmness and other mechanical properties, overwhelm different properties as the coefficient of thermal expansion, conductivity and thermal transport [1]. The composites are lightweight materials with high strength to weight ratio.

The polymers, metals and biomaterials are the choices of materials for 3D printing technology. Bio-based plastics represent 1% of the 290 million tons of the plastics produced annually. The renewable resource-based bioplastics are vital for a sustainable and circular economy [2]. In the industrial, medical and agricultural sectors, the natural fibres based reinforcement is gaining prominence. Some of the commonly used natural materials for reinforcement in the composite are leaves of plants, grass, jute, coir and seeds. The advantages of using natural fibres are as follows; renewable, eco-friendly and poses no environmental or health hazards.

- The renewable nature of the fibres makes it an attractive solution for the many real-time applications in the fabrication of biocomposites and bioplastics, since it is abundant in nature.
- There is no health hazard in the usage of natural fibers and the fillers are employed along with the natural fibres to improve the mechanical properties and are used in biomedical field.
- There is no pollution in the usage of natural fibres and this makes it an optimum solution for the replacement of synthetic fibers in many applications. The synthetic fibres like glass and carbon are not completely biodegradable and eco-friendly.

The natural fibres in some cases are treated as waste and the utilization of this waste makes it a waste disposal solution. The natural fibre based polymer composites have good tensile strength and form a new generation of materials that replaces the classical materials in many applications. The natural fibres are mainly divided into three classes based on the part of the plants from which they are extracted; fibres extracted from the stem; fibres extracted from leaves and fibres extracted from fruits of plants [3]. The properties of natural fibres rely on the plant nature and the locality in which the plants are grown. The extraction technique and age also contribute to vital factors [4]. The coir is a rigid and hard multicellular fibre with a central portion called "lacuna". The sisal is one of the vital leaf fibres that have good strength. The oil palm fibres are tough and have similar properties like the coir fibres. The pineapple leaf fibre is soft and has high cellulose content. The pineapple leaf fibre is widely used as a reinforcing component in most of the composite materials due to its superior properties and cheap when compared to other natural fibres. The bio-composites, in general, are classified into two categories; green and partially eco-friendly composite. The green composite comprises of all the constituents that are renewable and the partially eco-friendly composites comprise of renewable resources and the man made materials. The performance of the natural fibres depends on the count of fibres, length, shape, arrangement and the bonding capability with the matrix element [5]. The bio-plastics are also gaining importance and have a wide number of applications in many fields. The biodegradable and non-biodegradable classes of bioplastics are there. In the classical home composting, a properly designed pit in the garden is enough for composting, however, bio-plastics requires industrial composting. The industrial composting is carried out at a high temperature and specific humidity levels. The first bioplastic was based on cellulose. The types of biocomposites are depicted in Fig. 7.1.

The green economy is characterized as economy that targets making issues of decreasing environment hazards, shortages and focuses on sustainable development without affecting the climate. The 2011 UNEP Green Economy Report contends "that to be green, an economy must not exclusively be effective, yet in addition reasonable. Reasonableness suggests perceiving worldwide and nation level value measurements, especially in guaranteeing a Just Transition to an economy that is low-carbon, asset productive, and socially comprehensive. The features of the sustainable green circular economy are as follows; the raw materials are taken from

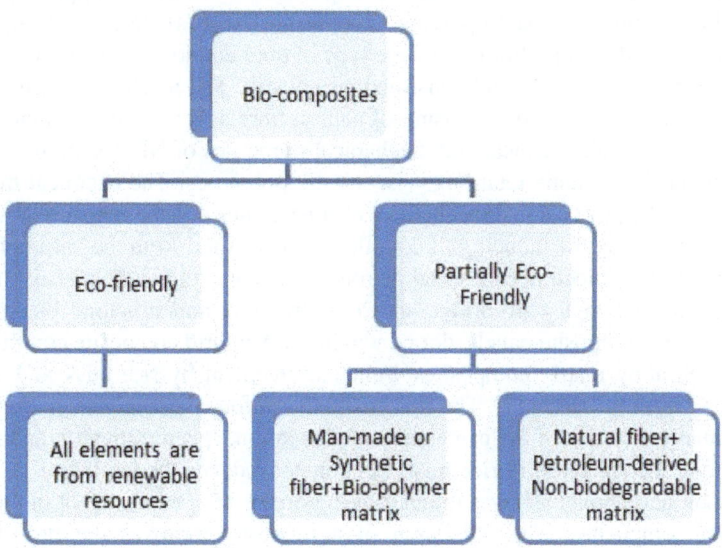

**Fig. 7.1** Types of bio composites [2]

nature and are renewable so that it will not perish and can be used continuously by future generations. The production of products will not generate any harm to the environment and customers using the product Minimization of waste and the improvement of waste management.

Section 7.2 highlights the typical biocomposites and its characteristics, Sect. 7.3 describes the typical bioplastics and its characteristics, Sect. 7.4 highlights the applications and finally, the conclusion is drawn in Sect. 7.5. This chapter focuses on the usage of agro waste as raw material for the fabrication of biocomposites and bioplastics, the applications of the bio composites and bioplastics are also discussed here.

## 7.2 Typical Bio-Composites and Its Characteristics

Rapid industrialization resulted in the development of industrially relevant synthetic polymers for a myriad of applications. Synthetic polymers are man made and are prepared using various polymerization techniques. The raw materials required for the preparation of these synthetic polymers are obtained from petroleum resources. With increased population resulted in increased consumption of fossil-fuel derived polymers and are now creating problems for the environment. Our natural resources such as air, water and soil are highly polluted because of the disposal of non-degradable plastics and other materials. Global warming and climate change-related issues also impart serious health problems and have posed serious risks of survival. This led to the push of finding suitable alternatives which are eco-friendly and sustainable and can be employed for demanding applications. The natural fibre reinforced based polymer composites (NFPCs) are gaining importance nowadays and the properties rely on the type of fibre employed. The NFPCs find its application in automobile and construction industry. The mechanical behaviour of the polymer was altered by the using of natural fibre as reinforcement material and in [6], the physical, chemical and mechanical properties of NFPCs are investigated with emphasis on flame retarding, viscoelastic properties. The chemical treatment can modify the physical and mechanical characteristics and the moisture absorption was minimized by the inducing of coupling agents. Apart from the automobile and industrial sector, biocomposites find its role in the construction field also; structural bicomposites examples are bridge and roof structure, non-structural bicomposites examples are composite panels, doors, windows. A hybrid composite comprising of Al–Mg–Si alloy matrix composites with reinforcement by rice husk and alumina $(Al_2O_3)$ was proposed in [7]. The rice husk ash reinforced composite gives superior performance in terms of the percentage of elongation, specific strength and fracture toughness. The increase in rice husk ash content reduces the hardness.

The oil palm boiler ash was found to be a good reinforcement agent in the epoxy polymer composites. The utilization of agro-waste improves the properties of composite and studies were carried out with different filler loadings and various particle sizes [8]. The ground nutshell/rice husk was used as the reinforcement agent in the hybrid bi-composite comprising of polypropylene. The ground nutshell/rice husk composite finds its application in false roofing in the construction field. The water absorption characteristics were 85% lower than the gypsum board and sound absorption capacity was also comparable with the gypsum boards [9]. A hybrid fibre composite comprising of Zea-coir polyester shows superior properties when compared with Zea polyester and polyester materials. The SEM analysis shows the efficient interaction between the Zea-coir fibre and the polyester. The composite material has excellent thermal stability and can be used as the replacement for the synthetic fibres [10].

The tensile properties of Betel nut husk (BNH) fibre were found to be proficient when compared with the kenaf and coir fibre. The thermal property of BNH fibre relies on the nature of its maturity like raw, ripe, and matured and is applicable as a reinforcement agent in the polymer composites [11]. In [12], the properties of kenaf-coir/PLA, bamboo-coir/PLA and kenaf-bamboo-coir/PLA composites were studied, the hybrid combination of kenaf-bamboo-coir with PLA was found to exhibit superior characteristics when compared with the others. The impact strength characteristics of coir polyester composites were studied in [8]; impact strength of 1.570 N-m and 1.275 N-m was observed in untreated coir polyester and treated composite materials [13]. The nanocrystalline cellulose (NCC) was extracted from the oil palm biomass waste and is used as the textile wastewater contaminant remediation [9]. The property of coir glass fibre hybrid fibre composite was varied by changing the fibre length and number of glass fibre mats [14]. The name suggests, biopolymers are environment-friendly biodegradable polymers having a natural or biological origin and that include polysaccharides (cellulose, starch, alginate, chitin etc.) and animal sources (silk, wool, gelatin etc.) [15]. Figure 7.2 depicts the types of biopolymers obtained from animal and vegetable waste.

Synthetic biodegradable polymers such as poly ε-caprolactone (PCL), polyvinyl alcohol (PVA), are also available in the market and are used in applications ranging from packaging to healthcare. The use of biopolymers in the global market is progressing by 17% over the forecast period from 2017 to 2021 and is expected to grow about 10 billion USD by 2021 [16]. Cellulose is the most abundant naturally occurring polymer and received much attention in the research community. They are biodegradable, non-toxic and biocompatible and are extensively used in fibres, wood, paper, cloth, cosmetic and pharmaceutical industries [17]. Cellulose obtained from citrus-peel waste is used for the preparation of zinc-impregnated nanocomposites sheets [17]. The authors characterized the nanocomposite by FT-IR, XRD, SEM and TGA and studied the antibacterial, anti-oxidant and photo-catalytic properties which might be useful biomedical and environmental applications. Nanocellulose is obtained from cellulose by mechanical or chemical methods. They are categorized into cellulose nanofibres (CNF's) and cellulose nanocrystals (CNC's). The acid hydrolysis method is the most popular chemical route for the preparation of nano cellulose. The preparation and processing of nano cellulose/ starch biopolymer nanocomposites are well documented in [18]. Cashew tree gum is a water-soluble polymer having the different composition of polysaccharide. Cashew tree gum-based nanocomposites with $Fe_3O_4$ nanoparticles were prepared

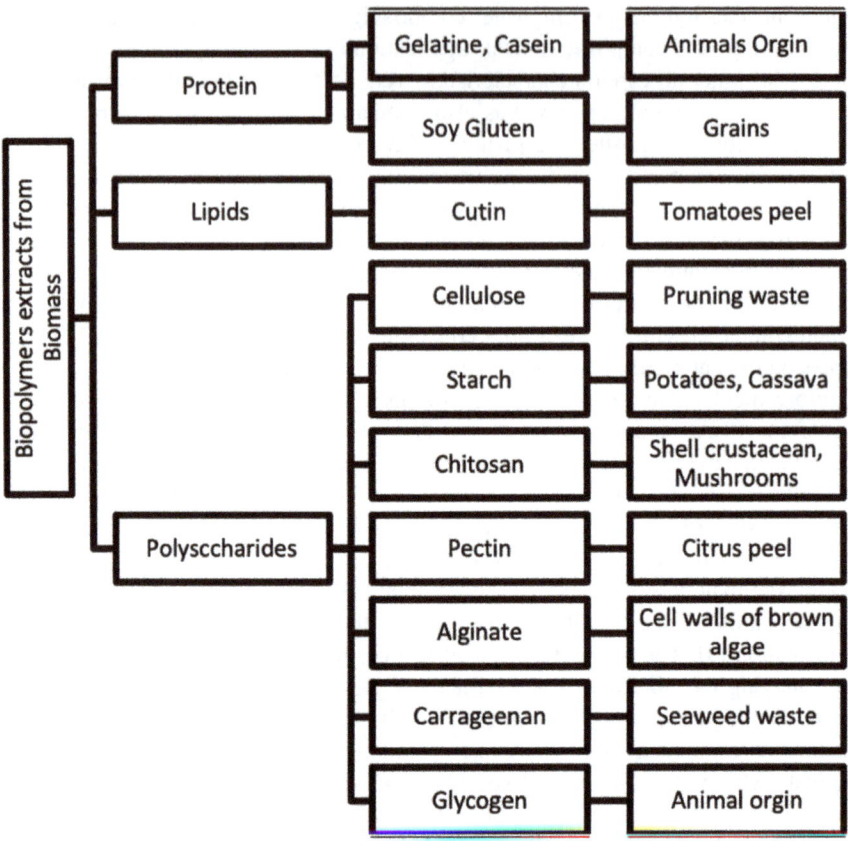

**Fig. 7.2** Types of biopolymers obtained from animal and vegetable wastes [2]

and characterized using different characterization techniques [19]. A new class of biopolymer nanocomposites were prepared by epoxidized soybean oil plastisized PLA with modified nanoclay (montmorillonite) [20]. The nanoclay was modified using fatty acid nitrogen containing compounds, which are synthesized from vegetable oil. Researchers used fatty amides (FA), fatty hydroxamic acids (FHA) and carbonyl difatty amides (CDFA) in their study. The nanocomposites were characterized using XRD, TEM and tensile measurements. The thermal analysis results showed improvement in thermal stability with the addition of fatty acid modified nano clay. The effect of the addition of fumed silica with different surface areas on thermo-mechanical and biodegradation properties of polylactic acid and poly ε-caprolactone based composites were investigated [21]. For biodegradation studies, the PLA and PCL composites were compression moulded (6 × 20 × 0.5 mm³). The samples were kept in contact with compost, which was made up using pruning residues, wood chips, dried residues and straw. The temperature of the compost medium was 40 °C and 58 °C respectively for PCL and PLA based nanocomposites. The relative humidity of about 50–70% was maintained. Forty samples of each specimen were vertically buried at 4–6 cm depth to ensure aerobic degradation conditions. The surface whitening in PLA composites indicated degradation happened and the whiteness was attributed due to water absorption and/or presence of products formed by hydrolysis; whereas the black colour on the surface of PCL specimens is expected to be due to the microbial growth [21]. Controlling the interface property of the biopolymer nanocomposite plays an important role in their usage in specific applications. The gas barrier properties of the nanocomposites containing PLA and nano cellulose particles functionalized by a hydrophobic lauryl chain was investigated [22]. The gas transport properties at different nano cellulose loadings were studied using gas-phase permeation technique and solvent desorption behaviour were studied using thermal desorption spectroscopy (TDS). Bone-inspired nanocomposites were prepared using soybean oil epoxidized acrylate and hydroxyapatite nanoparticles [23]. 3D finite element micromechanical models were carried out to investigate orthotropic responses of the nanocomposite filaments and the experimental data were compared with results from finite element and analytical micromechanical models. In [24], detection and characterization of nano clay in biopolymer nanocomposites were presented. A facile approach for the preparation of polyvinyl alcohol biopolymer nanocomposites with tailored mechanical property and exceptional antimicrobial property using cellulose nanocrystal (CNC) and graphene oxide (GO) was investigated [25]. The addition of CNC and GO favours the improvement in barrier properties towards water (25.7%) and water vapour (27.2%). Ag nanoparticles loading on PVA/CNC/GO composites exhibited significantly improved antibacterial property against Gram-negative (*E. coli*) and Gram-positive (*S. aureus*) bacteria. Poly (glycerol sebacate) (PGS) is a synthetic biopolymer is extensively used in the biomedical industry. Poly (glycerol sebacate) (PGS)/gelatin biopolymer nanocomposites were prepared using a solvent-free approach is reported [26], gelatin is used for copolymerization. The influence of nano clay and graphene oxide on the properties of nanocomposites such as hydrophilicity and degradation behaviour was investigated. The samples were characterized using FT-IR spectroscopy, XRD, dynamic-mechanical thermal analysis (DMTA), and TGA. The morphology of compounds is observed using scanning electron microscopy. Biopolymer nanocomposites membranes are used for the removal of metal ions from contaminated water. Chitosan has excellent heavy metal absorption capacity. The heavy metal ion removal capacity of polysulfone membranes containing CNT's/Chitosan was studied in [27]. The fabrication of biopolymer nanocomposites and porous scaffolds for tissue engineering applications are well explained in [15]. In short, the flexibility, biodegradability and biocompatibility of natural and synthetic biopolymer nanocomposites make it an attractive candidate for applications including packaging, water remediation, tissue engineering etc.

## 7.3 Typical Bioplastics and Its Characteristics

The bioplastic comprising of polymer, wood powder and corn powder was proposed in [11], the hybrid wood composite was prepared using the various components of corn powder. Figure 7.3 depicts some of the bio-based plastics manufacturing processes.

The bioplastics are environment friendly produced from renewable resources which is an alternative for petroleum-based plastics. The plastics are mostly used for packing in Foodservice, Agriculture, Consumer Electronics, Automobile and other consumer goods. The bio-based polymers are biocompatible and biodegradable. Natural polymers or bioplastics are organic or inorganic [28]. The biopolymers naturally obtained from rubber, nucleic acids, polypeptides, polysaccharides, etc. [29]. The increase in the rate of environmental awareness, consumer awareness, green feedstock, product quality which thrives for the bioplastics [30]. The processing of the bioplastics reduces the $CO_2$ emission 30–70% than conventional plastics [31]. The bioplastics have several advantages over conventional plastics, such as they are partially biodegradable, reduces the emission of greenhouse gases by reducing the use of petroleum-based products, manufacturing of bioplastic can be done with existing machines, and it is eco-friendly. It has a few disadvantages also such as cost and degrades when exposed to higher temperature, brittleness etc. [32]. Figure 7.4 depicts the bioplastics based on biodegradable and non-biodegradable materials.

The commonly used biodegradable bioplastic is the thermoplastic starch (TPS) which shares about 60%, Polylactic Acid (PLA) shares 20%, Cellulose Acetate (CA) shares 15% and other types bioplastics shares 5% of global

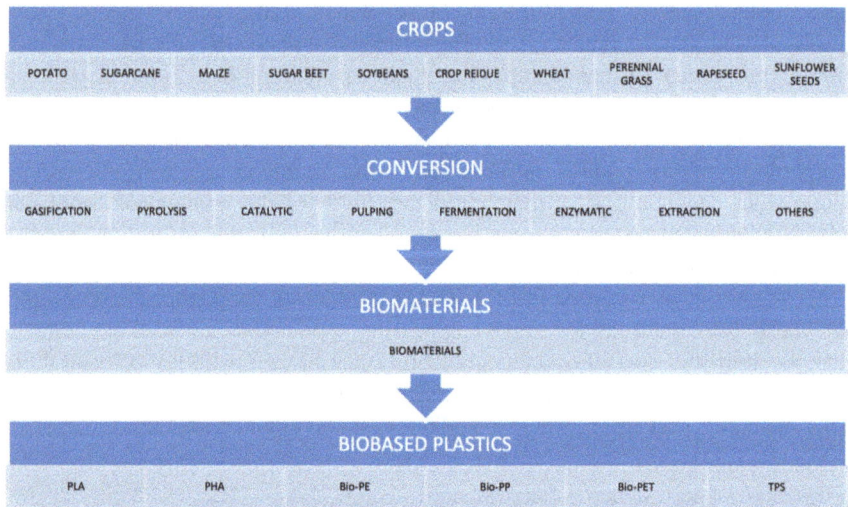

**Fig. 7.3** Bioplastics manufacturing techniques

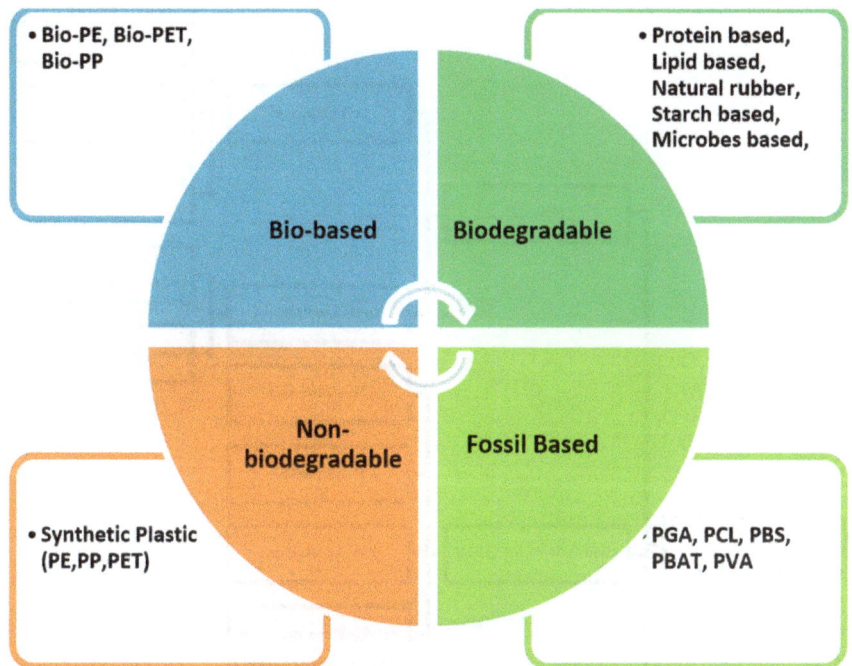

**Fig. 7.4** Types of bioplastics, both biodegradable and non-biodegradable

consumption. The demand for the bioplastics was increasing with consumer concern about the environment [31]. The starch is one of the major sources from the plants for the production of bioplastics. Genetically modified plants are directly used for bioplastic production [33]. The thermoplastic starch is obtained from cane sugar for the manufacture of bioplastic. The bacteria are a source for PHB, as it converts PHA to PHB by a metabolic process for the bioplastic production process [34]. The algae and seaweeds are the sources for bioplastic production. These are found to have abundant biomass and can be cultivated easily in the physical environment [35]. Figure 7.5 depicts the synthesis of different bioplastics.

The thermoplastic starch is produced by the extrusion of the starch along with shearing, temperature and plasticization to form a melted thermoplastic substance which is then transformed based on the application by thermoforming or injection moulding [36]. The PLA is produced from the renewable resources like fermented plant starch such as corn, cassava, sugarcane, sugar beet pulp etc., which by hydrolysis produce nutrient-rich hydrolysate then fermented to give Lactic acid. The lactic acid by downstream recovery and polymerization generates polylactic acid (PLA) [37]. The cellulose acetate is obtained from the Hemp/flax pulp of the harvested stalks by the process like activation of the pulp, then by acetylation to produce cellulose tri-acetate and again hydrolyzed to give the product Cellulose acetate (CA) [38]. The Table 7.1 depicts some of the typical bioplastics and the synthesis process.

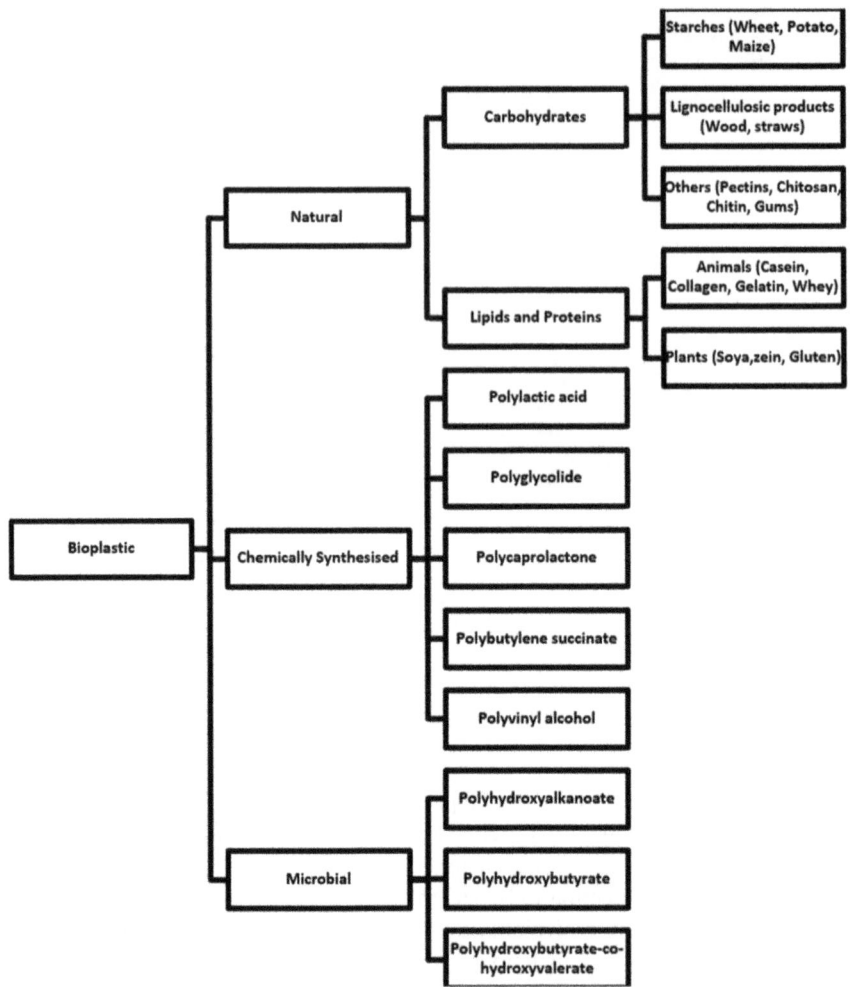

**Fig. 7.5** Bioplastics based on synthesis

## 7.4  Applications of Bio-Composites and Bioplastics

In [2], a detailed study has been made in the applications of natural fibres based composite materials. The agro-waste ash has high silica content and suits well as reinforcement material in the fabrication of aluminium matrix composites and gains importance in automobile and aerospace applications [7]. The results in [12] reveal that the hybrid combination of plant fibres can generate 100% plant fibre composite. The 100% green composite comprising of plant fibres is efficient and can be used rather than combining with the synthetic fibres. The untreated coir polyester composite has high impact strength when compared with the treated coir polyester

**Table 7.1** Application of different types of bioplastics

| Bioplastic | Raw material | Application | Reference |
|---|---|---|---|
| Bio-based PE (poly ethylene) | Crude oil, natural gas, or Shale gas, Methane | Plastic bags<br>Plastic films<br>Geomembranes and Containers bottles | [47] |
| Bio-based PP (Polypropylene) | Feedstock Methane | Automotive industry | [47] |
| Bio-based PET (thermoplastic polymer resin) | Feedstocks | Packaging<br>Bottles<br>Electronics<br>Textile | [47] |
| Bio-based PVC (Polyvinyl Chlorine) | Feedstocks | Inflatable products<br>Toys<br>Gloves<br>Tubing | [48] |
| Bio-based PC (Polycarbonates) | Feedstocks | Construction, Automobile and medical | [49, 50] |
| Bio-based PU (Polyurethanes) | Natural oil polyols | 3D printing<br>Regenerative medicine<br>Biomedical implants | [51] |
| Cellulose Acetate (Cellulose esters) | Cotton linters | Synthetic fibres<br>Cigarette filters | [52, 53] |
| Polylactic Acid (PLA) | Monomer lactic acid | Disposable cups | [54] |
| Polyhydroxyalkanoates (PHAs) | Agro and food wastes | Agriculture<br>Packing films<br>Bottles<br>Medical | [55] |
| Polybutylene Succinat (PBS) | Succinic acid and 1, 4-butanediol (BDO) | Disposable products<br>Packaging films<br>Drug encapsulation | [56] |
| Bio-based Polyethylene oxide (PEO) | Crude oil, Fossil resources | Industry and Medicine | [57, 58] |
| Bio-based Polyamide (PA) | Wool, silk, castor oil | Textiles<br>Transportation industry | [59] |
| Chitosan | Shells of shrimp and other crustaceans | Food packing, biomedical, agriculture | [60, 61] |
| Thermoplastic Starch (TPS) | Plants, Animals | Films, bags, pens, cutlery, and expanded packages | [62] |
| Cellophane | Wood, Cotton, or hemp | Plastic film | [63] |
| Polyesteramides (PEAs) | Petroleum product | Biomedical applications | [64] |
| Alginate | Algae | Films and coatings, packaging, paper, textile and wound dressing | [65, 66] |

(continued)

**Table 7.1** (continued)

| Bioplastic | Raw material | Application | Reference |
|---|---|---|---|
| Polytrimethylene Terephthalate (PTT) | Corn sugar | Carpet and textile fibres, Monofilaments, Films, Nonwoven fabrics | [67, 68] |
| Polyglycolic Acid (PGA) | Vegetable wastes | Sutures Biomedical applications | [69] |
| Polybutylene adipate-co-terephthalate (PBAT) | Fossil resources | Plastic bags Cling wrap, Paper cups, Food packing, Mulch film | [70] |

composite [13]. The agro-waste has the property of wastewater remediation and NCC extracted from oil palm removes waste dirt, waxy substances, hemicellulose and lignin in textile effluent [14]. The small size coir fibre exhibits good property as a reinforcement agent and lightweight property of hybrid composite material makes it suitable for engineering applications [39]. The natural fibre/agro waste hybrid composite material, the thermoplastic property makes it suitable for packaging and automotive applications [40]. The life cycle analysis (LCA) reveals that both the reinforcement and matrix material derived from agro-waste depicts a reduction in environmental pollutants and pave a way towards sustainable green economy [41]. The natural fibre composites play a significant role in construction applications [42, 43]. The hybrid natural fibre reinforced composite material comprising of sugarcane and coir find its applications in real-time due to high tensile and flexural strength [44]. Similarly, printed PLA preferred due to high tensile and less shrinkage with selection of optimal process parameters of FDM Machine [45, 46]. The Table 7.1 depicts the applications of different types of bioplastics.

## 7.5 Efficient Utilization of Agro Waste for a Sustainable Green Circular Economy

In India, 70% of the population is in rural areas and farming is the occupation of many people. Agro waste utilization empowers people in rural areas and improves their living conditions. As an example, after pineapple harvesting, the agro-waste generated can be used for the manufacturing of composite material. As an example, the pineapple is a major fruit crop in Kottayam district of Kerala and pineapple leaf fibre is widely used as a reinforcement agent in the composite material. For the efficient utilization of agro-waste, community empowerment project can be formulated. A typical triple helix model of the community empowerment project is depicted in Fig. 7.6. The business organization is depicted by B and is termed as community members. The second part in triple helix model in the university; research scholars from different domains can contribute fruitful solutions for the

**Fig. 7.6** Triple helix model of the community empowerment project

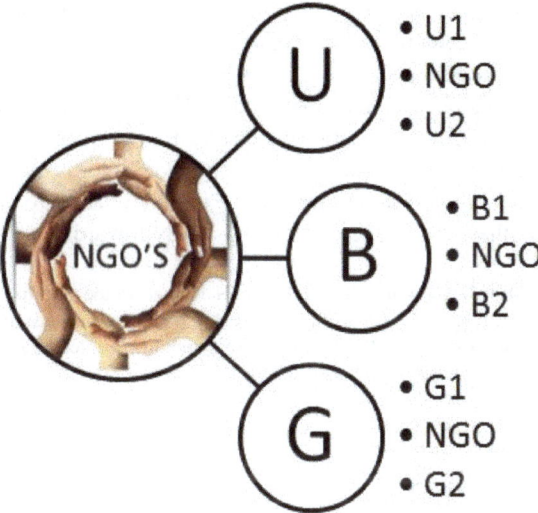

utilization of waste by applying appropriate technology. The NGO plays a vital role in establishing a link between rural area people and business organizations, government and university.

The agro-waste generated is dumped into a wasteland without utilizing it, the scientific knowledge can be employed to convert the agro-waste into useful products. The natural fibre-based composites and plastics can meet the socio-economic and environmental challenges thereby creating employment opportunities in rural areas, achieves the sustainable development goal formulated by the United Nations. This chapter focuses on the sustainable green circular economy. The features of the sustainable green circular economy are discussed in Sect. 7.1. The main raw material for the fabrication of biocomposites and bioplastics are renewable in nature and the manufacturing process does not generate environment pollutants. The utilization of agro waste is proposed in this work and it paves a way towards the prevention of waste and waste management.

## 7.6 Conclusion

The role of biocomposites and bioplastics in sustainable development are highlighted in this chapter. The properties of biocomposites and bioplastics rely on the treatment procedure and processes employed in the conversion of agro-waste. The manufacturing techniques and characteristics of typical biocomposites and bioplastics are stated in this chapter. The applications of the biocomposites and bioplastics are also highlighted in this work. The applications of biocomposites and bioplastics are immense in different sectors like the automotive industry, agriculture and medicine. The efficient utilization of agro-waste is a remedy for waste disposal and also empower the people in rural areas. The technology intervention is needed for the conversion of agro-waste into useful products thereby generating a sustainable circular green economy. The selection of appropriate fibre or agro-waste for a specific application needs effective decision making and requires technological intervention. Currently, the automotive sector is using natural fibres and agro-waste to a large extend and innovative research ideas are needed to utilize the agro-waste in other sectors thereby paving a way towards a sustainable green economy.

**Acknowledgements** The author S.N Kumar would like to acknowledge the CSIR SRTP 2020 programme since this work was a part of a summer internship. The author S.N Kumar would also like to acknowledge the support from Schmitt Center for Biomedical Instrumentation of Amal Jyothi College of Engineering.

# References

1. Zhang, C., Quirino, R.L., Sun, J.: Biobased polymers and composites. Int. J. Polym. Sci. (2018)
2. Reddy, M.M., Vivekanandhan, S., Misra, M., Bhatia, S.K., Mohanty, A.K.: Biobased plastics and bionanocomposites: current status and future opportunities. Prog. Polym. Sci. **38**(10–11), 1653–1689 (2013)
3. Mwaikambo, L.: Review of the history, properties and application of plant fibres. Afr. J. Sci. Technol. **7**(2), 121 (2006)
4. Gurunathan, T., Mohanty, S., Nayak, S.K.: A review of the recent developments in biocomposites based on natural fibres and their application perspectives. Compos. A Appl. Sci. Manuf. **1**(77), 1–25 (2015)
5. Ku, H., Wang, H., Pattarachaiyakoop, N., Trada, M.: A review on the tensile properties of natural fiber reinforced polymer composites. Compos. B Eng. **42**(4), 856–873 (2011)
6. Tong, F.S., Chin, S.C., Doh, S.I., Gimbun, J.: Natural fiber composites as potential external strengthening material—a review. Indian J. Sci. Technol. **10**(2), 1–5 (2017)
7. Alaneme, K.K., Akintunde, I.B., Olubambi, P.A., Adewale, T.M.: Fabrication characteristics and mechanical behaviour of rice husk ash—alumina reinforced Al–Mg–Si alloy matrix hybrid composites. J. Mater. Res. Technol. **2**(1), 60–7 (2013)
8. Rizal, S., Fizree, H.M., Hossain, M.S., Gopakumar, D.A., Ni, E.C., Khalil, H.A.: The role of silica-containing agro-industrial waste as reinforcement on physicochemical and thermal properties of polymer composites. Heliyon. **6**(3), e03550 (2020)
9. Guna, V., Ilangovan, M., Rather, M.H., Giridharan, B.V., Prajwal, B., Krishna, K.V., Venkatesh, K., Reddy, N.: Groundnut shell/rice husk agro-waste reinforced polypropylene hybrid biocomposites. J. Build. Eng. **27**, 100991 (2020)
10. Balaji, N.S., Chockalingam, S., Ashokraj, S., Simson, D., Jayabal, S.: Study of mechanical and thermal behaviours of zea-coir hybrid polyester composites. Mater. Today: Proc. **1**(27), 2048–2051 (2020)
11. Yusriah, L., Sapuan, S.M., Zainudin, E.S., Mariatti, M.: Characterization of physical, mechanical, thermal and morphological properties of agro-waste betel nut (Areca catechu) husk fibre. J. Clean. Prod. **1**(72), 174–180 (2014)
12. Yusoff, R.B., Takagi, H., Nakagaito, A.N.: Tensile and flexural properties of polylactic acid-based hybrid green composites reinforced by kenaf, bamboo and coir fibers. Ind. Crops Prod. **30**(94), 562–573 (2016)

13. Prasad, G.E., Gowda, B.K., Velmurugan, R.: A study on impact strength characteristics of coir polyester composites. Procedia Eng. **1**(173), 771–777 (2017)
14. Shanmugarajah, B., Chew, I.M., Mubarak, N.M., Choong, T.S., Yoo, C., Tan, K.: Valorization of palm oil agro-waste into cellulose biosorbents for highly effective textile effluent remediation. J. Clean. Prod. **10**(210), 697–709 (2019)
15. Okamoto, M., John, B.: Synthetic biopolymer nanocomposites for tissue engineering scaffolds. Prog. Polym. Sci. **38**(10–11), 1487–1503 (2013)
16. Mellinas, C., Ramos, M., Jiménez, A., Garrigós, M.C.: Recent trends in the use of pectin from agro-waste residues as a natural-based biopolymer for food packaging applications. Materials (Basel) **13** (3) (2020)
17. Ali, A., Ambreen, S., Maqbool, Q., Naz, S., Shams, M.F., Ahmad, M., Phull, A.R., Zia, M.: Zinc impregnated cellulose nanocomposites: synthesis, characterization and applications. J. Phys. Chem. Solids **98**, 174–182 (2016)
18. Ilyas, R.A., Sapuan, S.M., Norrrahim, M.N.F., Yasim-Anuar, T.A.T., Kadier, A., Kalil, M.S., Atikah, M.S.N., Ibrahim, R., Asrofi, M., Abral, H., Nazrin, A., Syafiq, R., Aisyah, H.A., Asyraf, M.R.M.: Nanocellulose/Starch Biopolymer Nanocomposites: Processing, Manufacturing, and Applications. Elsevier Inc. (2020)
19. Surya, K., Ramesan, M.T.: No title. Polym. Compos. **38** (S1), E66–E73 (2016)
20. Al-Mulla, E.A.J., Suhail, A.H., Aowda, S.A.: New biopolymer nanocomposites based on epoxidized soybean oil plasticized poly(lactic acid)/fatty nitrogen compounds modified clay: preparation and characterization. Ind. Crops Prod. **33**(1), 23–29 (2011)
21. Fukushima, K., Tabuani, D., Abbate, C., Arena, M., Rizzarelli, P.: Preparation, characterization and biodegradation of biopolymer nanocomposites based on fumed silica. Eur. Polym. J. **47**(2), 139–152 (2011)
22. Rigotti, D., Pegoretti, A., Miotello, A., Checchetto, R.: Interfaces in biopolymer nanocomposites: their role in the gas barrier properties and kinetics of residual solvent desorption. Appl. Surf. Sci. **507**(December 2019) (2020)
23. Bahmani, A., Comeau, P.A., Montesano, J., Willett, T.L.: Extrudable hydroxyapatite/plant oil-based biopolymer nanocomposites for biomedical applications: mechanical testing and modeling. Mater. Des. **174**, 107790 (2019)
24. Schmidt, B., Petersen, J.H., Bender Koch, C., Plackett, D., Johansen, N.R., Katiyar, V., Larsen, E.H.: Combining asymmetrical flow field-flow fractionation with light-scattering and inductively coupled plasma mass spectrometric detection for characterization of nanoclay used in biopolymer nanocomposites (2009)
25. Bai, H., Liang, Z., Wang, D., Guo, J., Zhang, S., Ma, P., Dong, W.: Biopolymer nanocomposites with customized mechanical property and exceptionally antibacterial performance. Compos. Sci. Technol. **199**, 108338 (2020)
26. Aghajan, M.H., Panahi-Sarmad, M., Alikarami, N., Shojaei, S., Saeidi, A., Khonakdar, H.A., Shahrousvan, M., Goodarzi, V.: Using solvent-free approach for preparing innovative biopolymer nanocomposites based on PGS/gelatin. Eur. Polym. J. **131**(April), 109720 (2020)
27. Refaat Alawady, A., Ali Alshahrani, A., Ali Aouak, T., Mohamed Alandis, N.: Polysulfone membranes with CNTs/Chitosan biopolymer nanocomposite as selective layer for remarkable heavy metal ions rejection capacity. Chem. Eng. J. **388**(January), 124267 (2020)
28. Mark, J.E., Allcock, H.R., West, R.: Inorganic Polymers. Prentice Hall, Englewood, N. J. (1992)
29. McNaught, D., Wilkinson, A. (eds.): Compendium of Chemical Terminology (the '"Gold Book"'), 2nd edn. Blackwell Scientific Publications, Oxford (1997)
30. Yates, M.R., Barlow, C.Y.: Conserv. Recycl. **78**, 54–66 (2013)
31. https://www.icis.com/explore/resources/news/2011/06/22/9471602/bioplastics-projects-set-to-prosper/
32. https://www.universiteitleiden.nl/en/research/research-output/science/cml-ia-characterisation-factors
33. Rahman, A., Miller, C.D.: Microalgae as a source of bioplastics. In: Algal Green Chemistry, pp. 121–138. Elsevier (2017)

34. Getachew, A., Woldesenbet, F.: Production of biodegradable plastic by polyhydroxybutyrate (PHB) accumulating bacteria using low cost agricultural waste material. BMC. Res. Notes **9**(1), 1–9 (2016)
35. Rajendran, N., Puppala, S., Sneha Raj, M., Ruth Angeeleena, B., Rajam, C.: Seaweeds can be a new source for bioplastics. J. Pharm. Res. **5**(3), 1476–1479 (2012)
36. Vilpoux, O., Averous, L.: Starch-based plastics. Technology, use and potentialities of Latin American starchy tubers, 521–53 (2004)
37. Rivero, C.P., Hu, Y., Kwan, T.H., Webb, C., Theodoropoulos, C., Daoud, W., Lin, C.S.: Bioplastics from solid waste. In: Current Developments in Biotechnology and Bioengineering, pp. 1–26. Elsevier (2017)
38. Harrison, I., Huttenhuis, P.J., Heesink, A.B., Enschede, P.T.: BIOCA—Biomass Streams to Produce Cellulose Acetate
39. Prajapati, P., Sharma, C., Rana, R.S.: Evaluation of mechanical properties of coir and glass fiber hybrid composites. Mater. Today: Proc. **5**(9), 19056–19062 (2018)
40. Shesan, O.J., Stephen, A.C., Chioma, A.G., Neerish, R., Rotimi, S.E.: Improving the mechanical properties of natural fiber composites for structural and biomedical applications. In: Renewable and Sustainable Composites. IntechOpen (2019)
41. Puglia, D., Sarasini, F., Santulli, C., Kenny, J.M.: Manufacturing of natural fiber/agrowaste based polymer composites. In: Green Biocomposites, pp. 125–147. Springer, Cham (2017)
42. Parveen, S., Rana, S., Fangueiro, R.: Natural fiber composites for structural applications. Mechanics of Nano. Micro Macro Compos. Struct. 1–2 (2012)
43. Sethunarayanan, R., Chockalingam, S., Ramanathan, R.: Natural fiber reinforced concrete. Transp. Res. Rec. **1226**, 57–60 (1989)
44. Arul, M., Sasikumar, K.S., Sambathkumar, M., Gukendran, R., Saravanan, N.: Mechanical and fracture study of hybrid natural fiber reinforced composite—coir and sugarcane leaf sheath. Mater. Today: Proc. (2020)
45. Sandhu, K., Singh, J.P., Singh, S.: Some investigations on the tensile strength of additively manufactured polylactic acid components. In: Advances in Materials Processing, pp. 221–230. Springer, Singapore (2020)
46. Sandhu, K., Singh, S., Prakash, C.: Analysis of angular shrinkage of fused filament fabricated poly-lactic-acid prints and its relationship with other process parameters. In: IOP Conference Series: Materials Science and Engineering, vol. 561, no. 1, p. 012058. IOP Publishing (2019)
47. Xxx Siracusa, V., Blanco, I.: Bio-Polyethylene (Bio-PE), bio-polypropylene (Bio-PP) and bio-poly (ethylene terephthalate) (Bio-PET): recent developments in bio-based polymers analogous to petroleum-derived ones for packaging and engineering applications. Polymers **12**(8), 1641 (2020)
48. Cheng, L., Wu, W., Meng, W., Xu, S., Han, H., Yu, Y., Qu, H., Xu, J.: Application of metallic phytates to poly (vinyl chloride) as efficient biobased phosphorous flame retardants. J. Appl. Polym. Sci. **135**(33), 46601 (2018)
49. Park, S.J., Lee, J.E., Lee, H.B., Park, J., Lee, N.K., Son, Y., Park, S.H.: 3D printing of bio-based polycarbonate and its potential applications in ecofriendly indoor manufacturing. Add. Manuf. **31**, 100974 (2020)
50. Kuczynski, J., Boday, D.J.: Bio-based materials for high-end electronics applications. Int. J. Sustain. Dev. World **19**(6), 557–563 (2012)
51. Hojabri, L., Kong, X., Narine, S.S.: Novel long chain unsaturated diisocyanate from fatty acid: synthesis, characterization, and application in bio-based polyurethane. J. Polym. Sci., Part A: Polym. Chem. **48**(15), 3302–3310 (2010)
52. Glasser WG. 6.: Prospects for future applications of cellulose acetate. In: Macromolecular Symposia, vol. 208, no. 1, pp. 371–394. Weinheim: WILEY-VCH Verlag (2004)
53. Fischer, S., Thümmler, K., Volkert, B., Hettrich, K., Schmidt, I., Fischer, K.: Properties and applications of cellulose acetate. In: Macromolecular Symposia, vol. 262, no. 1, pp. 89–96. Weinheim: WILEY-VCH Verlag (2008)

54. Jem, K.J., van der Pol, J.F., de Vos, S.: Microbial lactic acid, its polymer poly (lactic acid), and their industrial applications. In: Plastics from Bacteria, pp. 323–346. Springer, Berlin, Heidelberg (2010)

55. Zinn, M., Witholt, B., Egli, T.: Occurrence, synthesis and medical application of bacterial polyhydroxyalkanoate. Adv. Drug Deliv. Rev. **53**(1), 5–21 (2001)

56. Nazrin, A., Sapuan, S.M., Zuhri, M.Y., Ilyas, R.A., Syafiq, R., Sherwani, S.F.: Nanocellulose reinforced thermoplastic starch (TPS), polylactic acid (PLA), and polybutylene succinate (PBS) for food packaging applications. Front. Chem. **8** (2020)

57. Bedian, L., Villalba-Rodríguez, A.M., Hernández-Vargas, G., Parra-Saldivar, R., Iqbal, H.M.: Bio-based materials with novel characteristics for tissue engineering applications—a review. Int. J. Biol. Macromol. **1**(98), 837–846 (2017)

58. Yasarla, L.R., Ramarao, B.V.: Lignin removal from lignocellulosic hydrolyzates by flocculation with polyethylene oxide. J. Biobased Mater. Bioenergy **7**(6), 684–689 (2013)

59. Winnacker, M., Rieger, B.: Biobased polyamides: recent advances in basic and applied research. Macromol. Rapid Commun. **37**(17), 1391–1413 (2016)

60. Kumar, M.N.: A review of chitin and chitosan applications. React. Funct. Polym. **46**(1), 1–27 (2000)

61. Khor, E., Lim, L.Y.: Implantable applications of chitin and chitosan. Biomaterials **24**(13), 2339–2349 (2003)

62. Carvalho, A.J.: Starch: major sources, properties and applications as thermoplastic materials. In: Monomers, polymers and composites from renewable resources, pp. 321–342. Elsevier (2008)

63. Del Nobile, M.A., Fava, P., Piergiovanni, L.: Water transport properties of cellophane flexible films intended for food packaging applications. J. Food Eng. **53**(4), 295–300 (2002)

64. Ghosal, K., Latha, M.S., Thomas, S.: Poly (ester amides) (PEAs)–scaffold for tissue engineering applications. Eur. Polymer J. **1**(60), 58–68 (2014)

65. Sun, J., Tan, H.: Alginate-based biomaterials for regenerative medicine applications. Materials **6**(4), 1285–1309 (2013)

66. Pawar, S.N., Edgar, K.J.: Alginate derivatization: a review of chemistry, properties and applications. Biomaterials **33**(11), 3279–3305 (2012)

67. Hwo, C., Brown, H., Zhang, D., Sun, C.: Inventors, Shell Oil Co, assignee. Poly (trimethylene terephthalate) based meltblown nonwovens. United States patent application US 10/000,671 (2002)

68. Li, M., Wang, D., Xiao, R., Sun, G., Zhao, Q., Li, H.: A novel high flux poly (trimethylene terephthalate) nanofiber membrane for microfiltration media. Sep. Purif. Technol. **15**(116), 199–205 (2013)

69. Sun, X., Xu, C., Wu, G., Ye, Q., Wang, C.: Poly (lactic-co-glycolic acid): applications and future prospects for periodontal tissue regeneration. Polymers **9**(6), 189 (2017)

70. Ferreira, F.V., Cividanes, L.S., Gouveia, R.F., Lona, L.M.: An overview on properties and applications of PBAT based composites. Polym. Eng. Sci. **59**(s2), E7-E15 (2019)

# Chapter 8
# An Optimal Utilization of Waste Materials in Concrete to Enhance the Strength Property: An Experimental Approach and Possibility of 3D Printing Technology

**Meyyappan Palaniappan**

**Abstract** Concrete is a versatile composite material that consists of binding and filler materials with calculated water. Due to the sharp rising on the cost of these construction materials, the demands to search for various alternatives have increased. It is a challenge for the researcher's community to arrive at suitable alternatives without compromising the strength and durability aspects of conventional concrete. Many researchers observed that partial replacement of alternatives on the conventional materials can be suggested way to obtain a desired outcome instead of full replacement and all were in progress stage. The current scenario of the research in terms of arriving suitable alternatives has turned the attention towards the possible utilization of waste materials into the conventional concrete. In India, there was huge availability of waste materials and every year it is simply dumped into the site in an unutilized way. Here an attempt is made to effectively utilize some of the waste materials such as cow dung ash, rice husk ash, sugarcane bagasse ash, GGBS and marble dust into the concrete in the partial replacement in terms of volume fraction 0%, 10%, 20%, 30% and 40%. The study is focused to estimate the compressive strength of the concrete utilized with those waste materials. For this experimental study totally 126 cube specimens of size 150 mm × 150 mm were casted and water cured for 7 and 28 days. Compressive strength test results were tabulated and compared with various waste materials utilized. It is observed that the GGBS is found to be better among the other waste materials and the optimum ratio was identified as 40% replacement. In this paper, it is also discussed about the 3d printing technology which has drawn the attention of researchers in the world. The scope and possibility of 3d printing of concrete with the utilization of waste materials are highlighted in considering the views of future research and markets of construction sectors.

M. Palaniappan (✉)
Kalasalingam Academy of Research and Education, Tamil Nadu,
Krishnankoil 626 126, India
e-mail: meyyappan@klu.ac.in

© The Author(s), under exclusive license to Springer Nature Switzerland AG 2022
K. Sandhu et al. (eds.), *Sustainability for 3D Printing*, Springer Tracts
in Additive Manufacturing, https://doi.org/10.1007/978-3-030-75235-4_8

**Keywords** Concrete · Waste materials · Compressive strength · 3D printing

## 8.1  Introduction

One of the biggest challenge facing by the civil engineering industry is poor productivity of building materials due to rise in cost of raw materials, shortage of skilled labours, improper manufacturing methods, defective quality and over wastage [1, 2]. Due to the aforesaid issues, sometimes there will be a scarcity and raise in the cost of building materials for a good quality of building materials [3]. In order to bring down the cost, mostly quality is comprised and not it is not properly ensured about its standards. The other side, the effective disposal of waste materials in each and every sector is very difficult, since the recycling cost may adversely depend upon the more energy and high cost. So, in every process of manufacturing materials, the waste materials simply dumped into open land which lead to lots of pollution to the environment. Even the excessive use of cement will also leads of severe environmental problem like green house effect which may further damage the ozone depletion [4]. So, in the recent days, many of the research studies proved that, the waste materials can be utilized in a proper proportion into the production of building materials will address many of the issues in the construction sector and also safeguard the natural resources. It is an evident, that the locally available waste materials can effectively be introduced into the building material manufacturing process either in the partial or full replacement for cement or for the aggregates. Thereby, the cost of the building materials is cheap, efficient and durable which assures the sustainability era. The commonly used waste materials are fly ash, quarry dust, marble dust, granite dust, cow dung ash, rice husk, sugarcane bagasse ash, GGBS, demolished building materials, broken glasses, wood dust, plastics, red mud and silica fume etc. [5–7].

Many of the past researches, indicating the effectiveness of these waste materials into the concrete with a desired increase in the strength and durability properties. Here in this paper, an attempt is made to study the effectiveness on the compressive strength of the concrete with considering the waste material composition in a larger extent. For this study, the waste materials are turned out in the form of ashes and replaced for the cement in a partial replacement percentage of 10, 20, 30 and 40%. The main objective of this study is to evaluate the compressive strength of concrete in considering the incorporation of optimum percentage of waste materials such as cow dung ash (CD), rice husk ash (RH), sugarcane bagasse ash (SB), GGBS (GB) and marble dust (MD) for the age of 7 days and 28 days curing. In this paper, the wide scope and possibility of 3d printing of concrete is also given in addressing the important needs such as bringing down the cost, utilize the less number of skilled labours and maximize the reduction of wastage. The awareness on using the 3d printing technology has emerged and in mere future, it will have a large and wider implementation in addressing the eco-friendly/sustainable constructions with its additive manufacturing types.

## 8.2  Materials Used

The following concrete making materials are used in this study:

A. *Cement*

   53 grade Ordinary Portland Cement (OPC) confirming to IS: 8112 code standards. For the whole casting purposes of this study, the cement is obtaining in a single lot.

B. *Fine Aggregate*

   Locally available river sand passing through the IS sieve of 2.36 mm with a specific gravity of 2.63. It is confirming to zone II grading as per the codal provision IS: 382.

C. *Coarse Aggregate*

   Crushed granites stones are taken as coarse aggregates of size 20 mm with a specific gravity of 2.76 which confirms to the IS code 383.

D. *Water*

   Normal potable drinking water obtained from a single point of source is used for the entire study.

E. *Cow Dung Ash*

   The pulverized form of cow dung ashes are taken in this study which is passing through 300 microns sieve and the specific gravity is 2.58.

F. *Rice Husk Ash*

   In a nearby Srivilliputtur rice mill, rice husk is collected and burnt it to ashes. Its specific gravity is 2.16.

G. *Sugarcane Bagasse Ash*

   After juice extracted from sugarcane by squeezing process, the fibres are left out waste. In an open atmosphere, this is dried and burnt into ashes. The specific gravity is 2.38.

H. *GGBS*

   From one of the slag industries, it is obtained in a fineness white powder form. The specific gravity is 2.85.

I. *Marble dust*

   In Madurai, one of the marble shops, the marble powder is obtained in a fresh manner and its specific gravity is 3.02.

## 8.3  Experimental Investigation

In this experimental investigation, the mix design for M25 grade concrete was done based on the standards of IS: 10262:2009. The mix ratio arrived is 1 (Cement): 1.36 (Fine Aggregate): 2.68 (Coarse Aggregate) with 0.45 w/c ratio. The replacement proportions of waste materials in form of ashes such as cow dung ash, rice husk ash, sugarcane bagasse ash, GGBS and marble dust are 0, 10, 20, 30 and 40% for cement. In order to find the compressive strength property of concrete containing with and without replacement of above-mentioned proportions of waste materials. Cube specimens of size 150 mm × 150 mm × 150 mm is taken for this study. The mix ID and proportions for concrete constituent materials like cement, replacement waste materials, fine aggregate, coarse aggregate and water are listed in Table 8.1 in a detailed manner. For this study, 6 cube specimens are casted for conventional concrete and 24 specimens each for each waste materials replacement for 7 days and 28 days age of curing. So totally as 126 cube specimens are casted for this entire study. The details of specimens casted are listed below in Table 8.2. Initially, the concrete is added in a dry state of mix with the calculated amount of concrete ingredients such as cement, waste material ashes and fine aggregate and coarse aggregates thoroughly. Then the calculated amount of water is then added on the dry mixture and thoroughly mixed and filled in the cube moulds with three levels of compaction process. After a period of 24 hours, it is demoulded and made them to cure in the water curing tank for about the period of 7 days and 28 days.

With the help of compression testing machine (CTM), the loads are applied on the specimen. At the rupture point, the specimen gets completely failed and readings are noted down. The compressive strength of the specimen are calculated by the ratio of failure load to the cross-sectional area of the specimen. It is expressed in terms of N/mm$^2$.

## 8.4 Result and Discussion

Table 8.3 shows the experimental results of average compressive strength at the age of 7 days and 28 days for the various descriptions of concrete with and without the replacement of waste materials such as cow dung ash, rick husk ash, sugarcane bagasse ash, GGBS and marble dust as a partial replacement for cement in the proportions of 0, 10, 20, 30 and 40%. It is observed that, 28 days compressive strength is more than 7 days compressive strength and their age of curing days playing a vital role in enhancing the compressive strength of concrete. Therefore the age of curing days is directly proportional to the compressive strength of the concrete. This could be due to the formation of C–S–H gel w.r.t age of curing days. For this study, it is observed that, the 7 days strength attainment will be in the range of 63–70% of 28 days strength. From Table 8.3, it is noticed that, the minimum and maximum 28 days compressive strength will be 14.32 and 49.35 N/mm$^2$ for 40% replacement of cow dung ash and 30% replacement of GGBS in the cement. The conventional concrete (no replacement) has a compressive strength of 20.83 N/mm$^2$ and 31.84 N/mm$^2$ for the age of curing 7 days and 28 days respectively.

In Table 8.3, it is seen that the cow dung ash is replaced for cement into the concrete is showing the lowest compressive strength results among the other replacements. In comparison with the conventional concrete, the compressive strength is reduced by 5.02, 14.69, 29.99 and 55.02% for the replacement percentages of 10, 20, 30 and 40% replacement of cow dung ash. It is also observed

**Table 8.1** Details of Mix ID and its proportions

| S. No. | Mix ID | Replacement levels (%) | Cement (kg/m³) | CDA (kg/m³) | RHA (kg/m³) | SBA (kg/m³) | GGBS (kg/m³) | MD (kg/m³) | FA (kg/m³) | CA (kg/m³) | Water (litres) |
|---|---|---|---|---|---|---|---|---|---|---|---|
| 1 | CC | 0 | 424 | – | – | – | – | – | 576.6 | 1136.3 | 190.8 |
| 2 | CD10 | 10 | 381.6 | 42.4 | – | – | – | – | 576.6 | 1136.3 | 190.8 |
| 3 | CD20 | 20 | 339.2 | 84.8 | – | – | – | – | 576.6 | 1136.3 | 190.8 |
| 4 | CD30 | 30 | 296.8 | 127.2 | – | – | – | – | 576.6 | 1136.3 | 190.8 |
| 5 | CD40 | 40 | 254.4 | 169.6 | – | – | – | – | 576.6 | 1136.3 | 190.8 |
| 6 | RH10 | 10 | 381.6 | – | 42.4 | – | – | – | 576.6 | 1136.3 | 190.8 |
| 7 | RH20 | 20 | 339.2 | – | 84.8 | – | – | – | 576.6 | 1136.3 | 190.8 |
| 8 | RH30 | 30 | 296.8 | – | 127.2 | – | – | – | 576.6 | 1136.3 | 190.8 |
| 9 | RH40 | 40 | 254.4 | – | 169.6 | – | – | – | 576.6 | 1136.3 | 190.8 |
| 10 | SB10 | 10 | 381.6 | – | – | 42.4 | – | – | 576.6 | 1136.3 | 190.8 |
| 11 | SB20 | 20 | 339.2 | – | – | 84.8 | – | – | 576.6 | 1136.3 | 190.8 |
| 12 | SB30 | 30 | 296.8 | – | – | 127.2 | – | – | 576.6 | 1136.3 | 190.8 |
| 13 | SB40 | 40 | 254.4 | – | – | 169.6 | – | – | 576.6 | 1136.3 | 190.8 |
| 14 | GG10 | 10 | 381.6 | – | – | – | 42.4 | – | 576.6 | 1136.3 | 190.8 |
| 15 | GG20 | 20 | 339.2 | – | – | – | 84.8 | – | 576.6 | 1136.3 | 190.8 |
| 16 | GG30 | 30 | 296.8 | – | – | – | 127.2 | – | 576.6 | 1136.3 | 190.8 |
| 17 | GG40 | 40 | 254.4 | – | – | – | 169.6 | – | 576.6 | 1136.3 | 190.8 |
| 18 | MD10 | 10 | 381.6 | – | – | – | – | 42.4 | 576.6 | 1136.3 | 190.8 |
| 19 | MD20 | 20 | 339.2 | – | – | – | – | 84.8 | 576.6 | 1136.3 | 190.8 |
| 20 | MD30 | 30 | 296.8 | – | – | – | – | 127.2 | 576.6 | 1136.3 | 190.8 |
| 21 | MD40 | 40 | 254.4 | – | – | – | – | 169.6 | 576.6 | 1136.3 | 190.8 |

**Table 8.2** Specimen details

| Description of concrete | Replacement proportions (%) | No. of cube specimens |
|---|---|---|
| Conventional concrete | 0 | 6 |
| Concrete with cow dung ash | 10, 20, 30, 40 | 24 |
| Concrete with rice husk ash | 10, 20, 30, 40 | 24 |
| Concrete with Sugarcane bagasse ash | 10, 20, 30, 40 | 24 |
| Concrete with GGBS concrete | 10, 20, 30, 40 | 24 |
| Concrete with marble dust | 10, 20, 30, 40 | 24 |
| Total specimens | | 126 |

**Table 8.3** Experimental test results of compressive strength

| S. No. | Description of concrete | Specimen ID | Replacement (%) | Avg. compressive strength (N/mm$^2$) | |
|---|---|---|---|---|---|
| | | | | 7 days | 28 days |
| 1 | Conventional concrete | CC | 0 | 20.83 | 31.84 |
| 2 | Concrete with cow dung ash | CD10 | 10 | 20.09 | 30.24 |
| 3 | | CD20 | 20 | 19.13 | 27.16 |
| 4 | | CD30 | 30 | 14.13 | 22.29 |
| 5 | | CD40 | 40 | 9.89 | 14.32 |
| 6 | Concrete with rice husk ash | RH10 | 10 | 23.03 | 33.43 |
| 7 | | RH20 | 20 | 23.95 | 37.93 |
| 8 | | RH30 | 30 | 22.98 | 39.12 |
| 9 | | RH40 | 40 | 22.11 | 36.38 |
| 10 | Concrete with sugarcane bagasse ash | SB10 | 10 | 20.87 | 32.15 |
| 11 | | SB20 | 20 | 21.02 | 32.47 |
| 12 | | SB30 | 30 | 21.17 | 30.56 |
| 13 | | SB40 | 40 | 18.92 | 28.97 |
| 14 | Concrete with GGBS | GG10 | 10 | 25.47 | 39.84 |
| 15 | | GG20 | 20 | 28.71 | 43.38 |
| 16 | | GG30 | 30 | 32.86 | 49.35 |
| 17 | | GG40 | 40 | 29.34 | 44.62 |
| 18 | Concrete with marble dust | MD10 | 10 | 22.91 | 35.01 |
| 19 | | MD20 | 20 | 25.03 | 38.21 |
| 20 | | MD30 | 30 | 22.99 | 35.59 |
| 21 | | MD40 | 40 | 19.73 | 30.34 |

that, the compressive strength is in the declining trend, if the replacement percentage is increased because the cow dung ash has normally required more amount of water during mixing process and thereby the water will react in a slow manner in producing hydration of C–S–H gel. For the replacement percentage 30 and 40%,

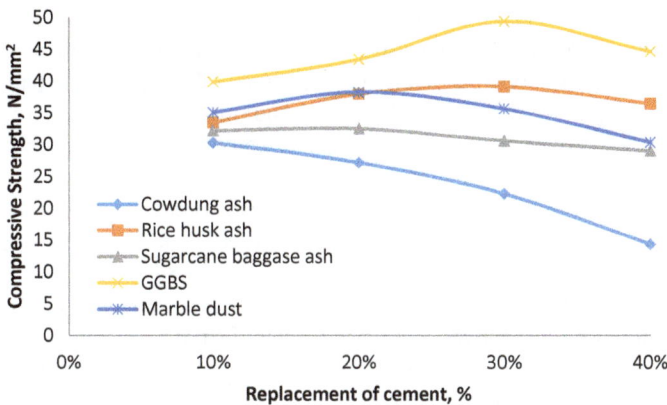

**Fig. 8.1** Compressive Strength at 28 days for different replacement of cement

there is a significant decrease in the compressive strength when compared to 10 and 20% replacement since it is not well bonding with the cement during mixing and hardening stages, if the volume of the cow dung ash increased. Based on the result, it is noticed that, 10% replacement of cow dung ash into the cement on concrete could be optimum in getting 95% of the compressive strength of conventional concrete.

In Fig. 8.1, it is understood that, rice husk ash replacement in the cement of the concrete has better performance in terms of enhanced compressive strength property than cow dung ash and sugarcane bagasse ash replacement. All the replacement percentage of rice husk ash has compressive strength more than the conventional concrete (31.84 N/mm$^2$). The replacement percentage of 10%, 20%, 30% and 40% of rice husk ash has the compressive strength more than 4.76%, 16.05%, 18.61% and 12.48% of conventional concrete respectively. This could be due to the content of amorphous silica present in the rice husk ash. The presence of this amorphous silica will improve the surface area of the transits zone of cement and aggregate boundary. Here, the compressive strength is in the high amplifying trend for 10–20% of replacement and for 20–30% replacement, the rise in compressive strength is very mild. (It is just 3.04% only). Further if the replacement to 40%, the compressive strength is decreased to 7.01%.

The sugarcane bagasse ash replacement in the cement content of the concrete shows the better performance than cow dung ash replacement but underrating performance than rice husk ash, marble dust and GGBS. In Fig. 8.1, it is understood that the sugarcane bagasse ash replacement shows very narrow marginal increase (less than 2%) for 10 and 20% replacement than conventional concrete but if the replacement percentage increased beyond 20%, the compressive strength found to be less. For 30% replacement, the compressive strength is reduced by 4.02%. Finally, for 40% replacement, the strength is further reduced by 5.21%. The optimum replacement proportion of sugarcane bagasse ash is arrived as 20%, Due to the presence of amorphous silica; it is exhibiting similar cementing properties even after the replacement is increased. For a maximum 40% of replacement, 90% of the conventional concrete strength is attained. The graphical pattern looks marginal increase up to 20% replacement and marginal decrease for 30–40% replacement. Moreover, the graph looks like a steady pattern whereas the cow dung ash shows a deep declining nature.

Concrete with the partial replacement by GGBS to cement shows an excellent performance in improving the strength properties than the other replacement materials. The compressive strength is found to be 39.84 N/mm$^2$, 43.38 N/mm$^2$, 49.35 N/mm$^2$ and 44.62 N/mm$^2$ for the replacement percentages of GGBS by 10%, 20%, 30% and 40% respectively. The percentage of strength increase is in the order of 20.10%, 26.61%, 35.49% and 28.64% for the replacement 10% to 40% respectively. The maximum compressive strength is almost 1.5 times the compressive strength for the replacement proportion 30% of GGBS in cement content. The graph as shown in Fig. 8.1 indicating that, the increasing trend in compressive strength up to 30% replacement of cement by GGBS and when it is increasing to 40%, there is a change in the trend (decreasing trend). This implies the reduction in the compressive strength by 10% but at the same time, this strength is more than the conventional concrete. Due to the fineness nature of GGBS, transformation of larger pores into finer pores is happened because of pozzolanic reaction and transformed state of calcium silicate hydrate gel into the concrete.

The replacement of cement with marble dust, produced some significant impact on the strength property of concrete. Its performance is higher than cow dung ash and sugarcane bagasse ash replacement. For 10% and 20% replacement of marble dust, the strength is higher than rice husk ash and if increased further beyond the attained strength is lower than rice husk ash. Except 40% replacement of marble dust, all the replacements had strength higher than conventional concrete. The compressive strength of concrete is increased by 9.06%, 16.68% and 10.54% for the replacement percentage 10%, 20% and 30% respectively in comparison with conventional concrete. Since it is a filler material and it will occupy the pores and makes the concrete denser. Once the pores are addressed, then strength will be increased. For 40% replacement, the strength is reduced by 4.71%. Some sudden slip for 30% and 40% of replacement by marble dust on the cement for the concrete is seen clearly in the Fig. 8.1, Since the volume of filler materials are more than the binding materials.

## 8.5  Scope and Possibility of 3D Printing of Concrete

3d printing of concrete is a computer-controlled production process/technique which creates three dimensional solid parts through an additive layer by layer process. Generally, the manufacturing process will be either additive or subtractive type. The conventional method of construction is based on subtractive type but the problem is the consumption of more energy on subtracting process and more waste obtained. In order to reduce the energy and materials cost, a new automation technology called as 3d printing of concrete is preferred more in the recent times of construction industry. This could be more advantageous for the construction activities of structures which are having complex geometries in which the usage of such expensive formworks can be neglected. The general principles adopted in 3d printing is modelling phase, printing phase and finishing phase. In modelling phase, the 3d image of a required item is created using CAD software program. Then in printing phase in order to print the machine reads the design through an.idle file format and sent that information to the printer and in final finishing phase, the printer forms that concerned item by depositing materials in layers by layers. It will start from the bottom layer to top layer w.r.t the shape and size. There are three methods of 3d printing work that will perform. They are selective laser sintering, stereo lithography and fusion deposition modelling methods. Mostly the 3d printing of concrete or construction is mainly based on fusion modelling method due to the precious and accuracy level of the work.

The scope of 3d printing can be established in recent days, since the construction industry is already experiencing the practice of CAD/CAM, this 3d printing will be user friendly if the requirements are correctly pointed out. The recent need of building information modelling (BIM) will assist the wide publicity of 3d printing to a greater extent. The major advantages of 3d printing in construction will be lowering of labour cost, production of less wastages, reduce the interference of humans in risky situations/sites, save time and effort. The construction experts and professionals can interact with the clients in a more efficient communicable way with available alternatives in the area of design and materials. The client will also get a clear and better solution for their doubted misconception in the construction field. Now the entire construction industry is focussing towards sustainability approach, there are many possibilities/opportunities for 3d printing of the construction of eco-friendly houses with the available organic, sustainable and eco-friendly building materials. These materials are available on the effective utilized way of waste materials. With this concept, much affordable, durable construction is possible in the path of 3d printing technology. This technology will also have the present hopeful opportunities towards a greener and most cost-effective trending situations in the mere future.

## 8.6  Conclusion

Based on this experimental study, the following conclusions are arrived:

1. The attainment of 7 days compressive strength will be in the range of 63–70% of 28 days strength for the concrete with and without consideration of waste materials.
2. The replacement of cow dung ash into the cement by 10% could be found as the optimum level in getting 95% of the compressive strength of the conventional

concrete. Beyond these replacement levels, the compressive strength gets a drastic reduction.

3. It is observed that, around 18.61% increase in compressive strength, if the rice husk ash is utilized in the concrete. Therefore the optimum replacement percentage of rice husk ash is found to be 30% and beyond that, the trend is in decline nature.

4. For sugarcane bagasse ash, the compressive strength seems to be in very marginal increase up to 20% replacement and after that marginal decrease in the pattern seen.

5. The compressive strength is increased to almost 1.5 times of the conventional concrete for the replacement proportion 30% of GGBS in cement content and beyond that the strength decreases but that even will be more than the conventional concrete strength.

6. Some sudden slip for 30 and 40% of replacement by marble dust on the cement is seen. This observation indicates, the optimum replacement levels are up to 20% in enhancing the strength property.

7. In this paper, the scope and possibility of 3d printing are given in a detailed manner. This technology has a lot of scope and opportunities in the construction industry in addressing sustainable environmental needs and cost effecting trends in the nearby future.

# References

1. Bosela, P., Delatte, N., Obrati, R., Patel, A.: Fresh and G=hardened properties of paving concrete with steel slag aggregates. In: Proceedings, 9th International Conference on Concrete Pavements, San Francisco, California (2008)

2. Meyyappan, P.L., Kumaran, K., Gopalakrishnan, M., Harikrishnan, E.: Effect of glass fibers, flyash and quarry ash on strength and durability aspects of concrete—an experimental study. In: IOP Conference Series: Material Science and Engineering (2018). https://doi.org/10.1088/1757-899X/396/1/012001

3. Kaur, M., Kaur, M.: A review on utilization of coconut shell as coarse aggregates in mass concrete. Int. J. Appl. Eng. Res. 7(11), 7–9 (2012)

4. Olanipekun, E.A., Olusola, K.O., Ata, O.: A comparative study of concrete properties using coconut shell and palm kernel shell as coarse aggregates. Build. Environ. 41(3), 297–301 (2006). https://doi.org/10.1016/j.buildenv.2005.01.029

5. Halicka, A., Ogrodnik, P., Zegardlo, B.: Using ceramic sanitary ware waste as concrete aggregate. Constr. Build.Mater. 48(May), 295–305 (2013)

6. Ajnavi, S.: Bioconversion of cellulosic agricultural wastes. Masters Technology Dissertation, Department of Biotechnology Environment Science, Thapar University, 60601011 (2008)

7. Meyyappan, P.L., Amuthakannan, P., Sutharsan, R., Ahamed Azik Ali, M.: Utilization of M-sand and basalt fiber in concrete: an experimental study on strength and durability properties. In: IOP Conference Series: Material Science and Engineering (2019). https://doi.org/10.1088/1757-899X/561/1/012035

# Chapter 9
# 3D Printing Incorporated with Supply Chain Management and Associated Waste Production

**Gurcharan Singh Bhalla, Harmanpreet Singh, and Puneet Bawa**

**Abstract** Since the development of the additive manufacturing (AM) process also prominently known as 3D printing or rapid prototyping there is an exponential increase in its applications under various domains. 3D printing when incorporated with supply chain management can be really helpful to streamline the processes along with waste management. Various factors can be kept in mind while implementing 3D printing along with supply chain management. The waste that is generated from different manufacturing processes when they were being turned into final products can also be reduced to a great extent or can be eliminated if production is done by 3D printing. Rapid Prototyping (RP) is a layer-by-layer manufacturing process. Likewise, Computer-assisted design (CAD) can specifically be used to produce such tri-dimensional physical models. This manufacturing method gives engineers and designers an absolute ability to print the tri-dimensions layout of their concepts and models. Processes for RP includes a quick and cheap alternative for prototyping functional models in contrast with the traditional component production. The benefit of constructing a component layer-by-layer is that even the complex shapes can be easily made which though were the almost impossible to manufacture by machining process. RP can construct complex structures within structures, internal sections, and very thin-walled features equally quickly to construct a simple cube. AM technology emerges as an easy sell in the market to create complex shapes with the material needed and to enhance the design and simulation of complex structures. This results in disruption of technologies that have a global impact on the supply chain and the logistics of the business. The essence of this

G. S. Bhalla
Department of Production Engineering, National Institute of Technology,
Tiruchirappalli, Tamil Nadu 620015, India

H. Singh (✉)
Department of Mechanical Engineering, Thapar Institute of Engineering and Technology,
Patiala, Punjab 147004, India
e-mail: hsingh2_me16@thapar.edu

P. Bawa
Centre of Excellence for Speech and Multimodal Laboratory, Chitkara University Institute
of Engineering & Technology, Chitkara University, Chandigarh, Punjab 140401, India

© The Author(s), under exclusive license to Springer Nature Switzerland AG 2022      159
K. Sandhu et al. (eds.), *Sustainability for 3D Printing*, Springer Tracts
in Additive Manufacturing, https://doi.org/10.1007/978-3-030-75235-4_9

technology is the potential to deliver goods closer to client standards worldwide while maintaining the automated delivery of those products in real-time. It has major advantages over the management of the supply chain by reducing product, transport, and warehouse capital investment, and by encouraging stores to evaluate a global change in supply chain management. The primary goal is to acquire knowledge about the use and role of 3D printers in the management of the supply chain and to explore the consequences of AM for the management of the supply chain. The key goal of the research is to gain information on the use and role of 3D printing in supply chain management and to study AM's effect on supply chain administration.

**Keywords** 3D printing · Supply chain management · Waste management

## 9.1   Introduction

In the 1980s for the construction of models and parts of designs, the first sort of 3D layer-by-layer development Rapid prototyping method using CAD was established. Quick prototyping has been one of the earliest additive manufacturing techniques (AM). Among the significant advances presented by this method, the time and cost savings for product growth, human interaction is of main importance along with the ability to generate nearly any shape that may be a cumbersome task for manual machines [1].

With the assistance of RP, users can easily create and construct prototypes, for theoretical understanding and research, evaluate models. Figure 9.1 shows the various steps that are involved in product development using the rapid prototyping technique. At present, rapid prototyping also known as 3D printing technologies is not only limited to the production of prototypes, rather with the advantages of various materials and processes it is possible to manufacture finished products [2–6]. These innovations are also known as 3D printing, but they all stem from rapid prototyping techniques [4, 6]. In addition, it is important to note that the permissive parent technology of the rapid prototyping systems is other innovations CAD, computer-aided manufacturing (CAM), and computer numeric control (CNC).

Rapid prototyping is still in its developing phase and can be unsuitable for various tasks, where the commonly used manufacturing methods (like CNC) are needed to be employed. There can be instances where the dimensions of the parts are larger and it is unsuitable to print the products. The speed of prototyping materials is still an issue as fast and large-scale production by 3D printing is a challenge. It is obvious that ceramics and metals can at least be printed out, but not all widely used industrial materials. A description of the different production processes for additives is provided in Fig. 9.2. These processes should be divided between a liquid, solid base, a powder as focused in this adapted figure aforementioned. This analysis considers the procedures included as most effective in the past and promising for the future of the industry.

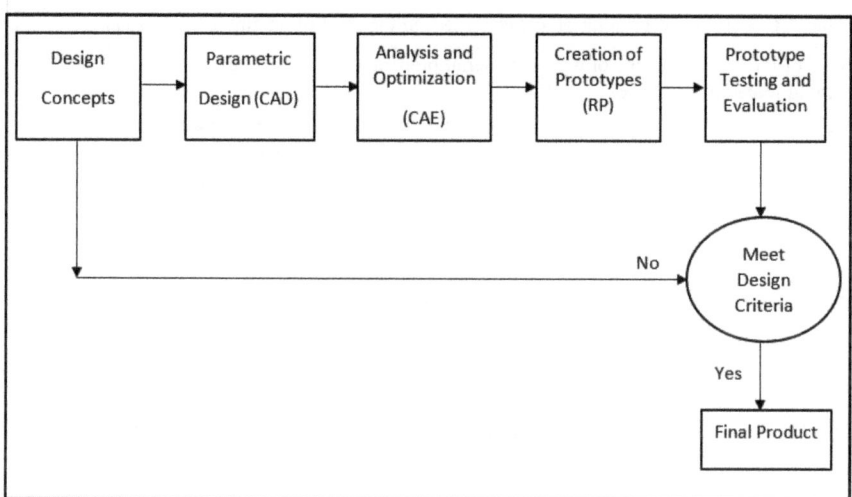

**Fig. 9.1** Block diagram showcasing the iterative steps for the cycle of product development [4]

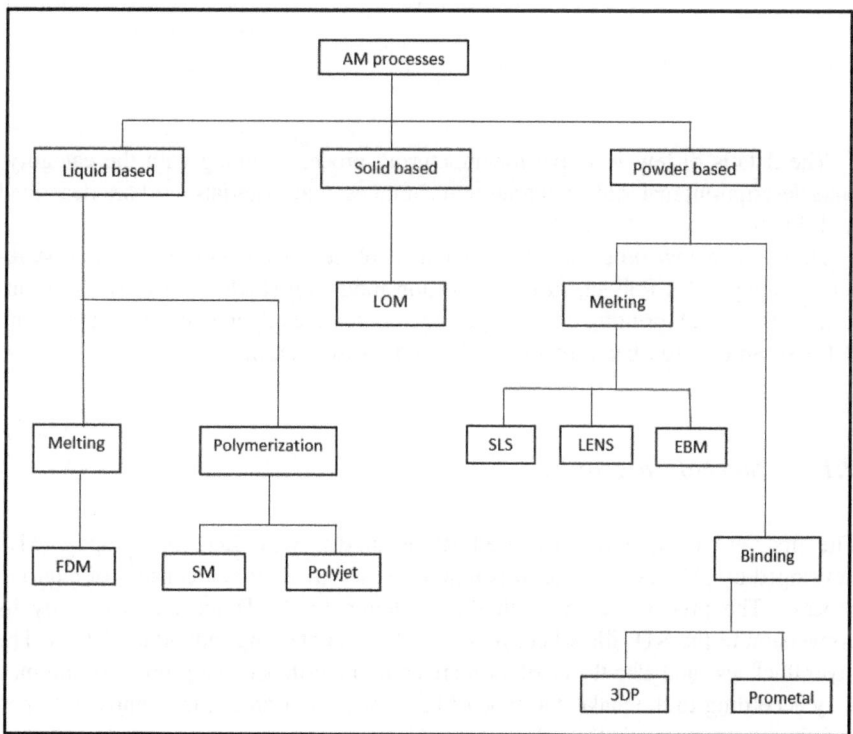

**Fig. 9.2** Three-dimensional printing processes [8]

**Table 9.1** Various 3d printing processes based on the state of starting materials [7]

| State of the starting material | Name of the Process | Material preparation | Materials |
|---|---|---|---|
| Liquid | SLA | The liquid resin in a Vat | UV Curable resin, ceramic suspension |
| | RFP | Liquid droplet in a nozzle | UV curable acrylic plastic, wax |
| | MJM | The liquid polymer in jet | Water |
| Filament/Paste | FDM | Filament melted in a nozzle | Thermoplastics, waxes |
| | Robocasting | Paste in nozzle | Ceramic paste |
| | FEF | Paste in nozzle | Ceramic paste |
| Powder | 3DP | Powder in bed | Polymer, metal, ceramic, other powders |
| | LENS | Powder injection through a nozzle | Metal |
| | EBM | Powder in bed | Metal |
| | SLM | Powder in bed | Metal |
| | SLS | Powder in bed | Thermoplastics, waxes, metal powder, ceramic powder |
| Solid Sheet | LOM | Laser cutting | Paper, plastic, metal |

The details of few Additive manufacturing processes along with the category-wise description, materials preparation techniques, and materials used are described in Table 9.1.

However, a few processes like laminated object manufacturing (LOM), stereolithography (SL), Polyjet, fused deposition modeling (FDM), selective laser sintering (SLS) 3D printing (3DP), pro metal, laminated engineered net shaping (LENS), an electron beam melting (EBM) are discussed in detail.

### 9.1.1 Stereolithography

The first and most frequently used RP technique was Stereolithography (SL), developed by 3D Systems, Inc. It is a liquid-based method that consists of a curing process. The process begins with the modeling by CAD tools followed by its conversion to the STL file where parts are sliced containing individual details. The layer thickness and also the resolution are equipment-dependent properties and may vary according to the make and type of equipment. A-frame is frequently designed to support structures that overlap.

When applied to the resin, the UV laser accurately solidifies for a sheet thickness according to the slices produced by the CAD design. After the completion of one

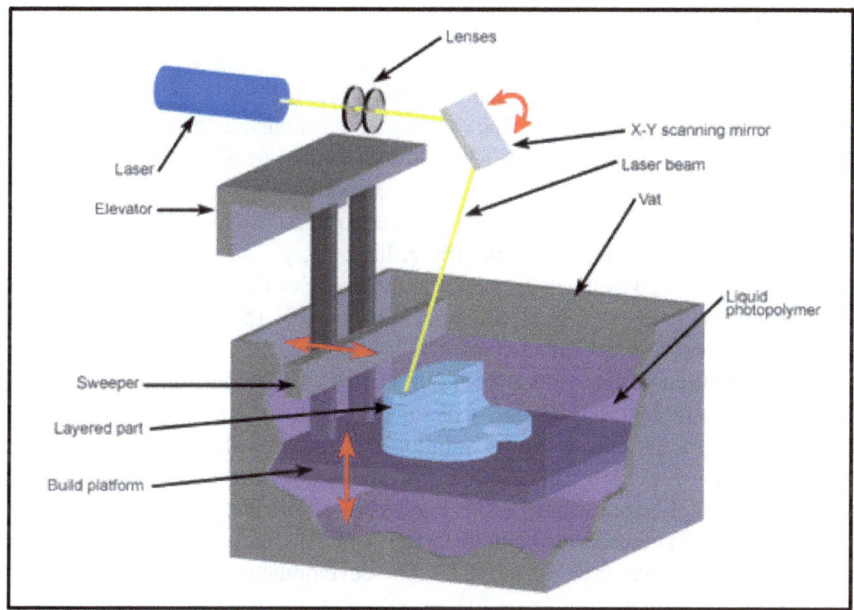

**Fig. 9.3**  Stereolithography [11]

layer the platform is lowered [2–4, 8]. The method is achieved by draining the waste and can be reused. One of its newer versions is also known as micro stereolithography. Using this method, we can achieve the thickness fewer than 10 μm [9, 10]. Figure 9.3 displays the stereolithography machine along with its basic parts.

The working principle behind this approach is ultraviolet curing or photopolymerization. This process transforms the liquid monomer or polymer into a solid polymer by ultraviolet light which helps in catalyzing the reaction. Powders can also be suspended in liquids such as ceramics [12]. There is a certain kind of errors that can be observed in the final part of the stereolithography process. One being an over-curing, which happened due to the overhanging sections as the fuse with the bottom layer is absent. The other one is the shape of the scanned line which gets changed due to the liquid's high viscosity further altering the thickness of the coating variable and making an error in the boundary position control. Lastly, the problem related to surface finishing arises as few of the tasks are performed manually [13]. These shortcomings can be resolved by using high-end equipment. There is a probability of using various materials when constructing a single product, and the process is known as multiple material stereolithography. For printing the product with different layers every time, the previous liquid resin needs to be drained out and the next resin should be poured and the process goes on in continuation till the final product is printed. There is need of a software for the scheduling process which is needed extensively to manage the different layers [14].

### 9.1.2  STL File

The STL file also known by the name Regular Tessellation language, was developed in 1987 by 3D Systems Inc. during the primary introduction of stereolithography to the market, due to which it was termed as STL. Though there are other file formats but the STL is a kid of standard file used for any Additive Manufacturing process. When the file is converted to STL the changes that may be seen are converted to the header, the triplet list of x, y, and z coordinates, and/or small triangles as well as the usual vector to the triangles from continuous geometry that was originally created in CAD application [4, 10, 15]. The internal and the external surfaces here are defined by the law of the right hand where vertices between a point and a line cannot be differentiated. Additional edges are added as the figure is removed. The slicing technique also introduces imprecision in the product to be printed due to the substitution of contour with discrete steps in the stairs by the algorithm [15]. In addition, the z-direction measurements should have several layer thicknesses values [13]. Figure 9.4 shows the location for the formation of the STL file corresponding to the data flow as in the process of rapid prototyping. The data flow in the STL file development program is shown in Fig. 9.5.

### 9.1.3  3DP

The 3DP process liquid binder is supplied regularly to the starch-based powder and by the use of roller, the layer-by-layer completion is done. The piston bed gets lowered on completion of each layer. The moment of the ink print head depends upon the part geometry made by the CAD tool. The method is quite similar to the

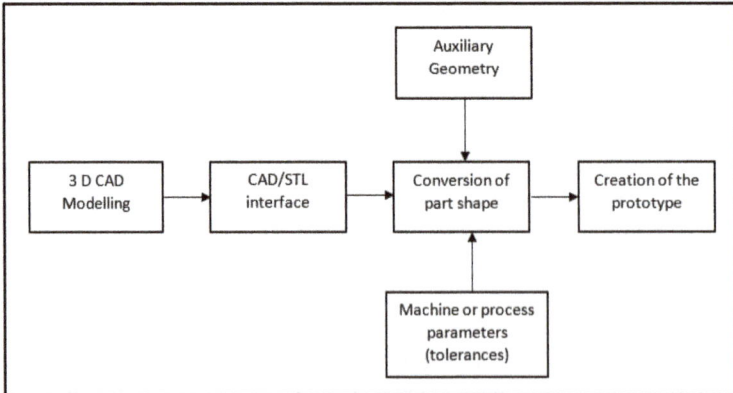

**Fig. 9.4**  Flow of data in rapid prototyping [4]

**Fig. 9.5** Data flow in STL file creation [4]

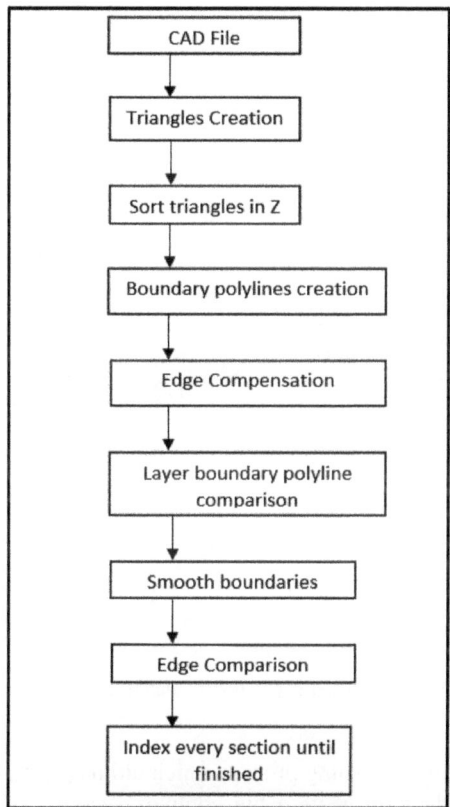

paper printing so it is named as 3DP as here the powder is lied on the bed which is glued by using the binder supplied through the print head. This method is capable of handling a wide range of polymers [2, 3, 10].

### 9.1.4   Fused Deposition Modeling (FDM)

The process of Fused deposition modeling (FDM) extrudes the filament supplied to it through the printing head after melting. The layer thickness that is printed on the base is of around 0.25 mm. The most commonly used filament materials are polyphenylsulfone (PPSF), polycarbonate (PC), PC-ISO, acrylonitrile butadiene styrene (ABS), and PC-ABS mixes, a medicinal-grade PC. The key advantage of this process is that no chemicals, curing resins, cheaper systems ,and the materials resulting from this are required, resulting in a more economical manner [2–4, 6]. The downside here is that the precision on the z-axis is weakly relative to the various additive manufacturing processes (0.25 mm) because there is always a need

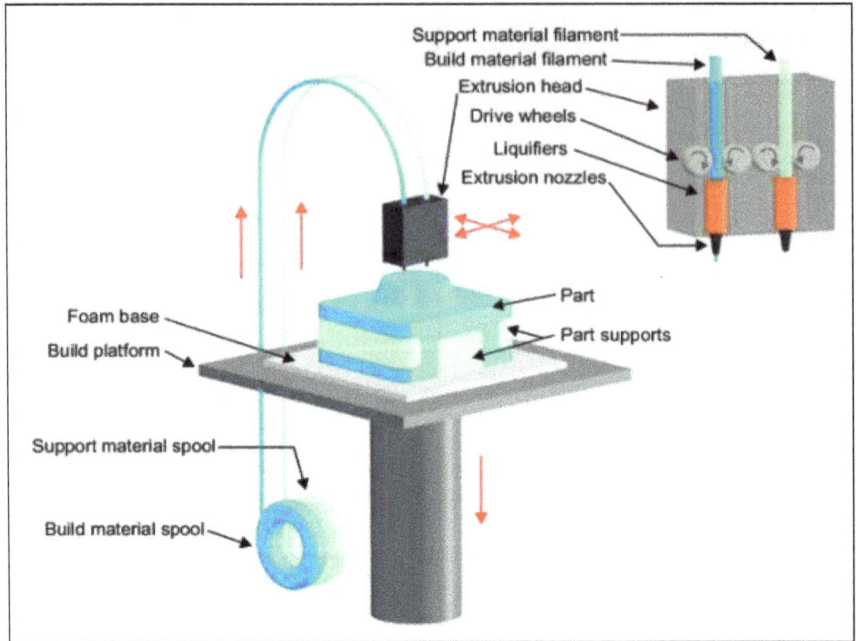

**Fig. 9.6** Fused deposition modeling [17]

for a finishing process which ultimately makes the process slow and that often takes days to create a big complex section [16]. The basic fundamentals of the fused deposition modeling process are shown in Fig. 9.6.

In this, a spool of thermoplastic wire as discussed above is supplied continuously through the nozzle, which melts the wire and also extrudes on the printing base. The moment of the extruder in the x, y, z direction is controlled by the computer according to the CAD Design. The Viscous strands (semi-liquid) are placed such that cross section of the prototype gets produced and following which the whole product is printed. In the newer version of machines, the supporting wax is supplied by another nozzle which is attached alongside the primary nozzle. This supporting structure is for the overhanging designs and can be easily removed in the later stages.

### 9.1.5 Prometal

Prometal is a tri-dimensional printing process used to make dye as well as injection machines. It is an inox stainless steel powder process. This same printing procedure happens as a liquid binder is separated through a layer of steel in jets. The powder is placed in a powder bed controlled by the installation of pistons which lower the bed after completion of each sheet and a piston supplying the material for each sheet. After finishing the process, the remaining powder must be gathered and discarded. Sintering, infiltration, and finishing processes are expected when a functional element is constructed [2, 3, 8]. The part is heated to 350 °F during the sintering process for 24-h hardening of the steel with such a 60% porous specimen. The component is infused with bronze powder in the course of integration, as it is heated together at over 2000 °F in 60% stainless steel and 40% bronze alloy [18–20].

## 9.1.6   Selective Laser Sintering

It is the 3-dimensional printing process where the powder which is loosely placed on the print base is sintered using the carbon dioxide laser beam. The moment of the laser beam is computer-controlled and depends on the design requirements of the product. The process basically consists of two chambers, the powder is supplied by the roller in a fixed amount for each consecutive layer. The powder which is not sintered act as the support structure to the complex shapes to be made. For instance, polymers also used are styrene-acrylic and polyamide (nylon), which have the same mechanical characteristics as injected parts [21, 22]. Composites or strengthened polymers, i.e., polyamide with fiberglass, can also be used in this process. They could even be fortified with metals such as copper. A binder is essential for metals. This may be a silicone binder, which would later be dissolved by the heating process or ground by making the change in the melting point of metals [21–23]. High strength aluminum pieces can be manufactured from polyvinyl alcohol, which is an organic binder [2, 3]. The key benefits of this approach are the large variety of materials that can be used. As mentioned earlier, the powder which is not sintered can be recycled and used in the printing of the next product. The drawbacks here are that the precision is constrained by the size of the substance samples, the oxidation must be stopped by conducting the procedure in an inert gas environment, and that the process takes place at a steady temperature close to the melting point. The Fig. 9.7 depicts the parts of the process.

## 9.1.7   Electron Beam Welding

Related to SLS is the process of electron beam melting (EBM). This is a relatively new form, but quickly is becoming more popular. In order to eliminate the possibility of metallic elements being imprinted as seen in Fig. 9.8, laser beams ranging from 30 to 60 kV are used for the melting of the powder into the vacuum chamber. In comparison, the process is very similar to the SLS mechanism.

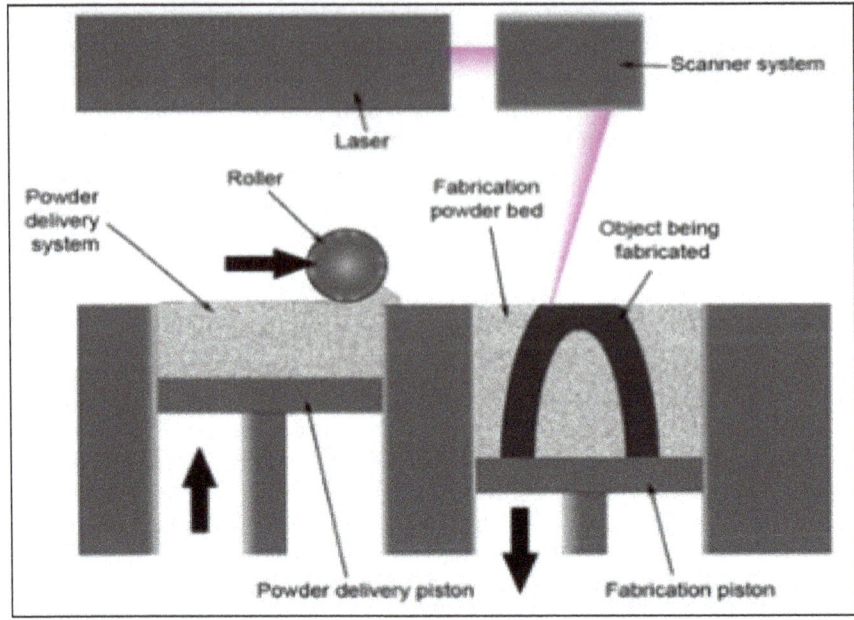

**Fig. 9.7** Selective laser sintering [24]

## 9.1.8 Laser Engineered Net Shaping

In this process of additive manufacturing, the metal powder is injected from the
nozzle which also has the laser beam, they are collaborated such that the metal
powder becomes liquid and is placed on the substrate as per the requirement of the
design of the product. The nozzle movement is controlled by the computer. The
whole process is performed in an inert atmosphere to avoid the oxidation problem.
The most commonly used materials as a powder are metals and a mixture of them,
such as nickel-based alloys, stainless steel, tooling steel, titanium-6 aluminum-4
vanadium, copper alloys, and much more. One of the problems is that erratic
heating and cooling processes could produce the residual tension, which can be
critical for high-precision processes, such as turbine blade repair [2, 3, 10, 26–28].
The component is fabricated in the process as shown in Fig. 9.9.

## 9.1.9 Laminated Object Manufacturing

It is a method that incorporates both subtractive and additive methodologies to
create a layer-by-layer product. Here the material is supplied in the form of a sheet,

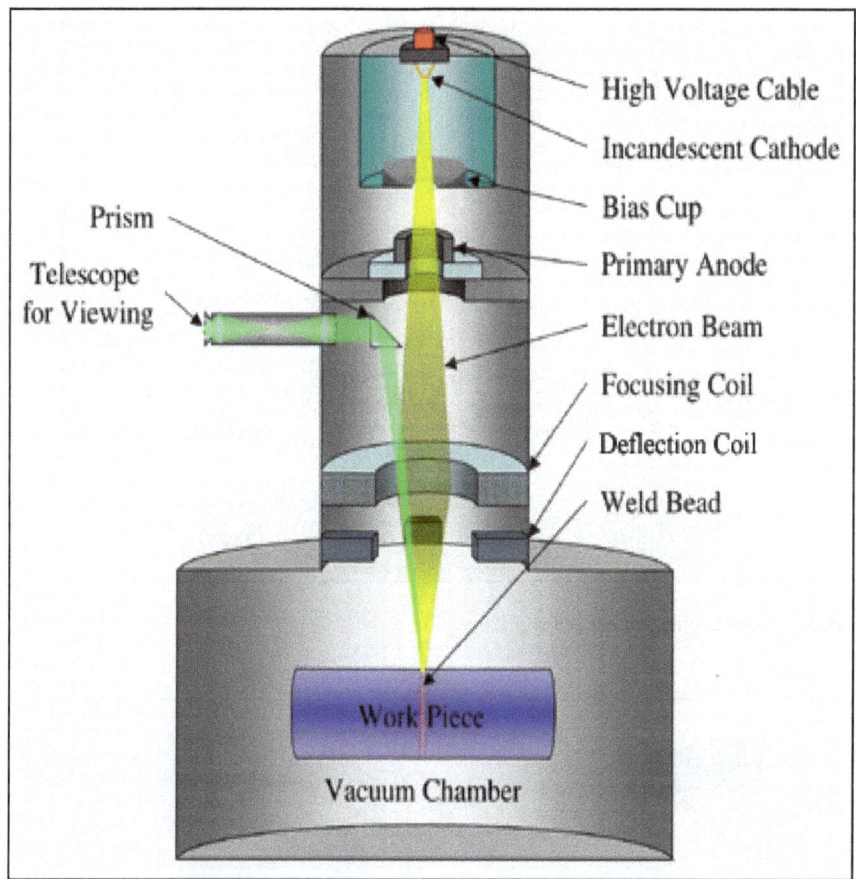

**Fig. 9.8** Electron beam welding [25]

which is cut by the carbon dioxide laser into proper requirement as per the design of the product. The layers are fused by the application of pressure and heat and by thermal adhesive coating. The benefits of this process are that they are relatively cheap, with no post-processing and supportive structures requirement, there are no deformation or change of step during the process, and the likelihood of constructing large pieces. The drawbacks here are the wastage of material, low surface definition, and the material is directionally dependent on machinability and mechanical characteristics, and the design of complex internal cavities is very difficult. This method can also be used for paper, plastic, and metals [2–4, 30] (Fig. 9.10).

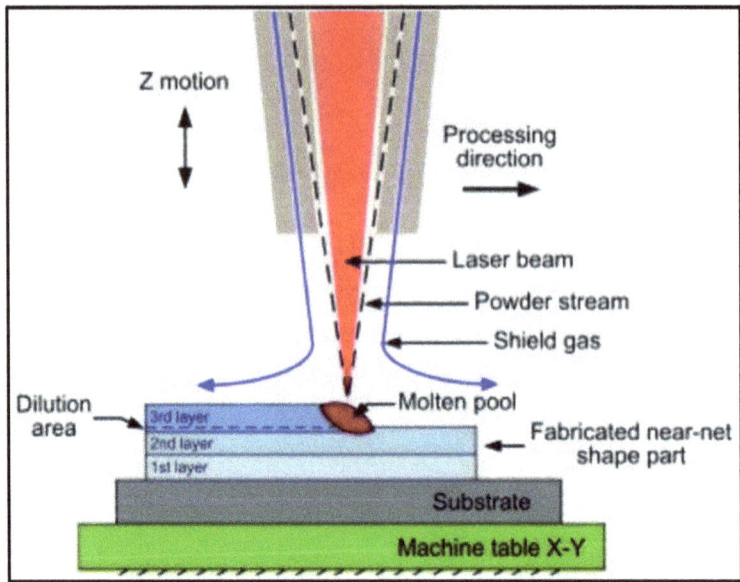

**Fig. 9.9** Laser engineered net shaping [29]

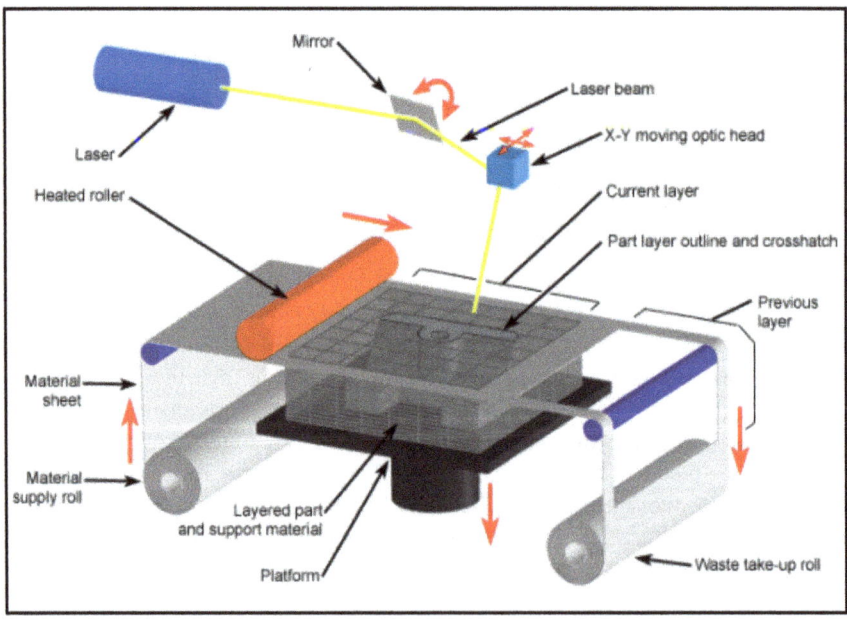

**Fig. 9.10** Laminated object manufacturing [31]

### 9.1.10  Polyjet

Here the inkjet technology to create physical versions, which travels in the x and y axes, depositing a photopolymer that is healed by ultraviolet lamps until each coating is finished. This method provides a high-resolution 16 μm thickness of the sheet. However, compared with stereolithography and selective laser sintering the sections produced by this process are of inferior quality. To support overhang features, a gel-type polymer is used and after the process, this liquid is tossed with water. This process will produce multiple color components [32–34].

## 9.2  Supply Chain Incorporated with 3D Printing

The supply chain is the network of businesses engaging in upstream and downstream partnerships by means of diverse processes and activities that produce demand such as products and services for consumers [35]. In other words, the supply chain constitutes an interconnected process where raw material is manufactured into the finished goods and then delivered to consumers based on their specifications as well as requirements [36]. The interest in management of the supply chain now involves the scheduling of the various supply chain members including suppliers, dealers, wholesalers, and retailers [37]. However, the growing focus has been paid to the efficiency, architecture, and study of the supply chain. The current interest tried to extend the definition of "reverse logistics" to include stock recovery for the purposes of recycling, replication, and reuse [38]. It focuses on issues such as supplier oversight, inventory management, purchasing, production management, customer support, planning facilities and, transportation and physical distribution information volatility [39]. It is important to enhance operating efficiency and better corporate goals, including enhanced quality, better customer experience, and increased profits [40]. Thus, through incorporating such networks into heuristic fields of both reverse and forward supply chains, the closed-loop supplier chain can be established [41]. In practice, the architecture, regulation, and function of the device need to be optimized to maximize the benefits of development over the product life cycle [42]. Work is currently ongoing on the appraisal of Supply Chain Management strategies at the triple bottom line, covering economic factors, environmental sustainability, and social responsibility. Increasing supply chain velocity leads to a faster reaction to business shifts or incidents that tend to increase the level of recovery from disruption [43]. The working phase concerning both the supply chain and the reverse logistics is represented in Fig. 9.11.

New advances in manufacturing, sourcing, and delivery have buffered the core fundamentals of the supply chain [45]. To understand these things, the types of goods aimed at enhancing environmental and social sustainability, 'sustainable growth,' which can be connected to their environmental and social values, shall be enforced [46]. Supply chain management is an assimilation of a market method that

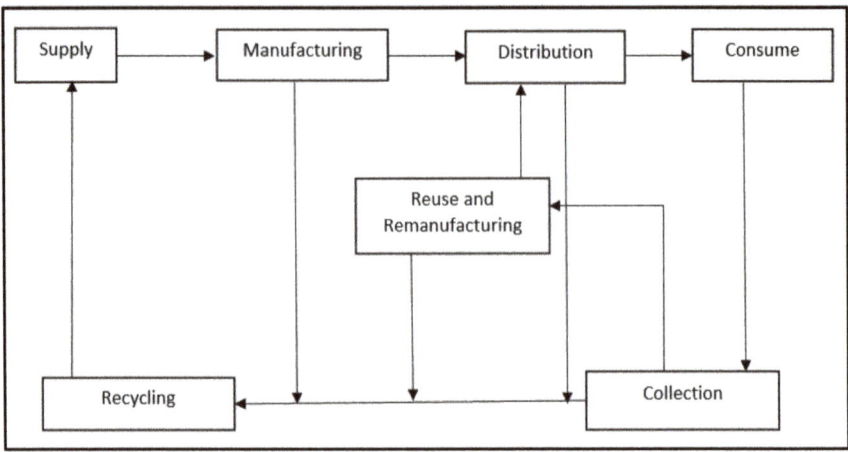

**Fig. 9.11** Block diagram showcasing schematic representation of the supply chain utilizing reverse logistics [44]

lists value to consumers over genuine vendors [47]. The core aspects of the supply chain can be determined through basic characteristics including inventory management strategy, tracking of expertise, overall cost engineering, operational philosophy, time plan, mutual sharing, supplier foundation, and flow of information [48]. It also involves the customer relationship management mechanism that upholds the jobs of the Manufacturer and Distribution Agreements (PSA) [49]. The other way to handle the supply chain is "green supply chain management" to overcome the link between the environment and the supply chain that has incorporated the green buying framework [50]. However, sustainable sources, better quality of the fuel, the elimination of transport vehicle emissions, safety of engine, rail, and airline industries are all obstacles to supply chain management [51]. There are huge sources of doubt as to whether supply chains are injurious, whether due to on-time productivity, average lateness and degree of variability, production instability as a result of process design, or system failures that have an effect on supply chain success [52]. Supply chain management depends on a virtual organization to support all partners in the delivery chain and on the two pillars of trust and teamwork to provide professional logistics for preference supply chain management [53].

The connection between site planning and Geography Information Systems has been established with the use of location models in the optimization suite that are developed by implementing solutions for the various programs [54]. It is one of the ways of incorporating any hierarchic inter-organizational environmental policy organization, and its main purpose is to support the green supply chain [55]. Management standards and efficiency of the supply chain are affected by quality control systems. This guarantees greater operational efficiency over upstream or downstream interactions [56]. Figure 9.12 indicates the description of the mechanism involved in the operation of the supply chain.

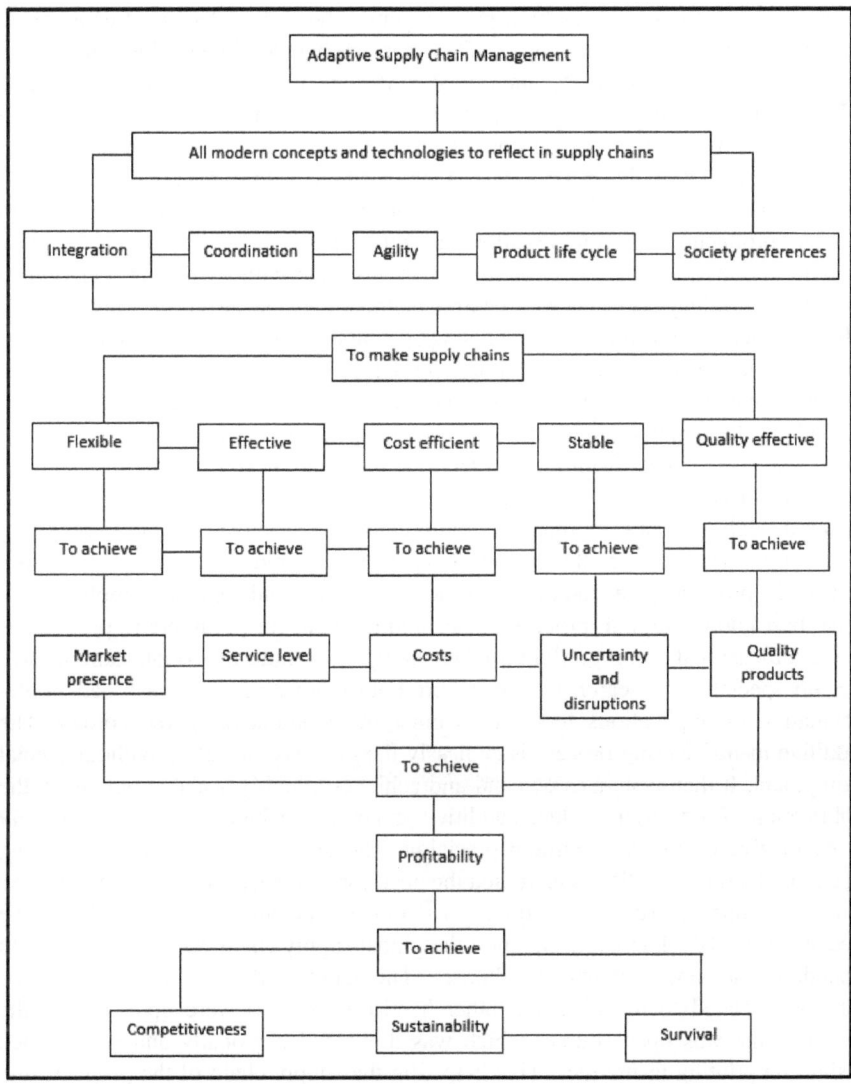

**Fig. 9.12** Adaptive supply chain management [57]

## 9.3   Impact of the 3D Printing on Supply Chain Management

For researchers and suppliers, the desired outcome of 3D printing in compliance with the supply chain management is a major obstacle. Optimism has contributed to the "industrial revolution" and has been viewed as a "gold rush" involving change in the supply chain techniques, competitiveness, and geography of manufacturing.

Various key factors, such as high print expense, the lack of technical experience, and the technological limitations of 3D printers are hampering the wide-ranging use of 3D printers. 3D printers' technological constraints hamper the widespread use of 3D printers. A conventional 3D printing service offers customers a way to fulfill their vision by using a special combination of design and manufacturing-related service. The biggest challenge is the resolution of the need to manufacture required components in superior efficiency at affordable prices that can be controlled through modern production technologies allowing supply chain management to produce any part in various positions and batch sizes at any time without paying attention to significant tooling costs. SCM's central methodology is the Food Supply Chain (FSC) which demonstrates complex supply chains consisting of multiple interdependent measures such as agriculture, food processing, storage, retail, and consumer management. On the other hand, the customer can use 3D printing to download the template and print it as a component, and also to print highly personalized components from the retailer. 3DP is economically competitive and has lower processing inputs and outputs of packaging in low volume markets and tailored high demand in manufacturing concerning aerospace and biomedical production. This reduces the consumption of energy, reduces capital expenditure across the product cycle, and transitions to more digital and localized supply chains. The best-known digital processing technique used for a different purpose is three-dimensional printing. More fundamentally, large industrial organizations have set up specialized research centers to promote innovation and the broad implementation of applications focused on emerging manufacturing technology. The additive manufacturing process is endlessly flexible, easy to set up without special equipment. It then generates demand and achieves a highly scalable output. In the fabrication of industrial products, additive processing techniques have been widely used to give customers optimum precision. The technical advancement enabling local production of 3DP has increased the need for ecological solutions and reduced emissions during urban revolution, smart cities, and other societal tools. In the recent COVID-19 Pandemic situation the global supply chain was disturbed due to lockdowns at industries and warehouses. The standard designs for the dedicated products related to the PPE and other hand-free devices were uploaded on the internet that was open source, which was downloaded globally and was printed using 3D printing techniques. This helped in the supply chain of the products and served as the better option in the pandemic situation.

## 9.4 Waste

The manufacturing or production of any commodity is always associated with wastage and scrap. Waste can be directly produced while the manufacturing operation like normal milling, drilling operation on the metal which produces the chips, or it can be found during the segregation done during the quality check of the final product produced. For making the process of the production more sustainable in terms of the economic and environmental point of view the wastage should be minimized, if produced should be made into reuse [58]. However, we can expect minimal wastage during the production process in 3D printing technology as compared to the conventional subtractive manufacturing operations.

### 9.4.1  Wastage in 3D Printing Processes

Wastage in 3D printing processes is though limited but it does exist in some form or the other even being an additive process. The overhanging parts when being printed need a special support structure that can be external or internal, which are removed during the post-processing of the final product [59–61]. Lack of adhesion in some additive processes also results in the inferior quality of products which are considered as scrap during the quality check. Solid sheet-based 3D printing process LOM (Laminated Object Manufacturing), can make the use of the wide variety of materials and can make the thin layers of the product. There is a continuous supply of the material sheet on the platform and a laser is used for the cutting of the sheet into the desired shape. The used sheet is continuously collected on the take-up roll. This process produces a lot of waste/scrap in the form of the remains of the sheet roll, which however can be re-casted into a sheet by recycling. Unpredictable shrinkages in the products made by the other 3D printing processes make the inferior products and thus the scrap is produced indirectly.

### 9.4.2  Waste Reduction in 3D Printing Processes

The wastage in any manufacturing process can only be minimized and controlled to some extent but cannot be completely vanished. The optimizing technique considering the printing path along with the orientation can significantly reduce the material consumption used for the support structure however the mechanical strength of the fabricated part cannot be assured alongside but can be improved by applying various strategies significantly [62, 63].

## 9.5  Conclusion

Additive Manufacturing techniques are known for constructing complex structures within structures, internal sections, and very thin-walled features equally quick to construct a simple cube. It has significant benefits over supply chain management through reducing inventory, transportation costs, and factory capital expenditure and warehouses providing the potential for assessing global supply chain management transformation. The primary goal of this analysis was to gain information about the use and involvement of 3D printing on supply chain management and researching AM's implications on supply chain management. 3D printing has been even used in the COVID-19 pandemic situations where the standard production and supply were halted. Few of the essential products were printed and supplied with ease using the open-source websites where the standard designs were uploaded. Wastage in the additive manufacturing process is few but still, the work to improve the waste/scrap reduction need to be done as a future scope.

# References

1. Ashley, S.: Rapid prototyping systems. Mech. Eng. **113**(4), 34 (1991)
2. Cooper, K.: Rapid Prototyping Technology. Marcel Dekker (2001)
3. Kochan, A.: Rapid growth for rapid prototyping. Assembly Autom. **17**(3), 215–217 (1997)
4. Noorani, R.: Rapid Prototyping Principles and Applications. John Wiley & Sons (2006)
5. Sandhu, K., Singh, G., Singh, S., Kumar, R., Prakash, C., Ramakrishna, S., Królczyk, G., Pruncu, C.I.: Surface characteristics of machined polystyrene with 3D printed thermoplastic tool. Materials **13**(12), 2729 (2020)
6. Wohlers, T.: Wohlers Report 2011. Wholers Associates (2011)
7. Guo, N., Leu, M.C.: Additive manufacturing: technology, applications and research needs. Front. Mech. Eng. 215–243 (2013)
8. Wohlers, T.: Wohlers Report 2009. Wholers Associates (2009)
9. Sandhu, K., Singh, S., Prakash, C.: Analysis of angular shrinkage of fused filament fabricated poly-lactic-acid prints and its relationship with other process parameters. In: IOP Conference Series: Materials Science and Engineering, vol. 561, no. 1, p. 012058. IOP Publishing (2019)
10. Phamand, D.T., Ji, C.: Design for stereolithography. In: Proceedings of the Institution of Mechanical Engineers, vol. 214, no. 5, pp. 635–640 (2000)
11. Online article on Stereolithography, by Custom Part. https://www.custompartnet.com/wu/stereolithography. Accessed on 30 Dec 2020
12. Kim, G.D., Oh, Y.T.: A benchmark study on rapid prototyping processes and machines: quantitative comparisons of mechanical properties, accuracy, roughness, speed, and material cost. Proc. Inst. Mech. Eng. **222**(2), 201–215 (2008)
13. Iancu, C., Iancu, D., Stamcioiu, A.: From cad model to 3D print via. STL file format. http://www.utgjiu.ro/revmec/mecanica/pdf/2010-01/13Catalin%20Iancu.pdf
14. Morvan, S., Hochsmann, R., Sakamoto, M.: ProMetal RCT(TM) process for fabrication of complex sand molds and sand cores. Rapid Prototyping **11**(2), 1–7 (2005)
15. Pro Metal R.C.T.: Pro Metal RCT rapid prototyping and digitals and casting services (2010) http://www.youtube.com/watch?v=Z8MaVaqNr3U
16. ExOne. 3D metal printing (2010). http://www.youtube.com/watch?v=i6Px6RSL9Ac&feature=related
17. Online article on Fused Deposition Modelling by Custom Part. https://www.custompartnet.com/wu/fused-deposition-modeling. Accessed on 30 Dec 2020
18. Lipke, D.W., Zhang, Y., Liu, Y., Church, B.C., Sandhage, K.H.: Near net-shape/net-dimension ZrC/W-based composites with complex geometries via rapid prototyping and displacive compensation of porosity. J. Eur. Ceram. Soc. **30**(11), 2265–2277 (2010)
19. Kruth, J.P., Mercelis, P., van Vaerenbergh, J., Froyen, L., Rombouts, M.: Binding mechanisms in selective laser sintering and selective laser melting. Rapid Prototyping J. **11**(1), 26–36 (2005)
20. Hwa-Hsing, T., Ming-Lu, C., Hsiao-Chuan, Y.: Slurry-based selective laser sintering of polymer-coated ceramic powders to fabricate high strength alumina parts. J. Eur. Ceram. Soc. **31**(8), 1383–1388 (2011)

21. Slavko, D., Matic, K.: Selective laser sintering of composite materials technologies. In: Annals of DAAAM and Proceedings, p. 1527 (2010)
22. Murr, L., Gaytan, S., Ramirez, D., et al.: Metal fabrication by additive manufacturing using laser and electron beam melting technologies. J. Mater. Sci. Technol. **28**(1), 1–14 (2012)
23. Semetay, C.: Laser engineered net shaping (LENS) modeling using welding simulation concepts. Pro Quest Dissertations and Theses, Lehigh University (2007)
24. Palermo, E.: What is selective laser sintering? Online article available on https://www. livescience.com/38862-selective-laser-sintering.html. Accessed on 30 Dec 2020
25. Article on Electron beam Welding: Principle, Working, Equipment's, Application, Advantages and Disadvantages. Available at https://www.mech4study.com/2017/04/electron-beam-welding-principle-working-equipment-application-advantages-and-disadvantages.html. Accessed on 30 Dec 2020
26. Liao, Y.S., Li, H.C., Chiu, Y.Y.: Study of laminated object manufacturing with separately applied heating and pressing. Int. J. Adv. Manuf. Technol. **27**(7–8), 703–707 (2006)
27. Vaupotic, B., Brezocnik, M., Balic, J.: Use of PolyJettechnology in manufacture of new product. J. Achievements Mater. Manuf. Eng. **18**(1–2), 319–322 (2006)
28. Singh, R.: Process capability study of polyjet printing for plastic components. J. Mech. Sci. Technol. **25**(4), 1011–1015 (2011)
29. Erosy, K., Celikl, B.B.: Utilization of additive manufacturing to produce tools. In: Design and Manufacturing. https://doi.org/10.5772/intechopen.89804
30. Petrovic, V., Vicente, J., Gonzalez, H., et al.: Additive layered manufacturing: sectors of industrial application shown through case studies. Int. J. Prod. Res. **49**(4), 1061–1079 (2011)
31. Online article on Laminated Object Manufacturing by Custom Part. https://www. custompartnet.com/wu/laminated-object-manufacturing. Accessed on 30 Dec 2020
32. Grimm, T.: User's Guide to Rapid Prototyping, Society of Manufacturing Engineers (2004)
33. Bletzinger, K.U., Ramm, E.: Structural optimization and form finding of light weight structures. Comput. Struct. **79**(22–25), 2053–2062 (2001)
34. Sweet Onions Creations. Architecture model and 3D printing—sweet onion creations (2007) http://www.youtube.com/watch?v=rEzugxybKmA
35. Mentzer, J.T., DeWitt, W., Keebler, J.S., Min, S., Nix, N.W., Smith, C.D., Zacharia, Z.G.: J. Bus. Logistics **22**(2), 1–25 (2001)
36. Beamon, B.M.: Int. J. Oper. Prod. Manage. **19**(3), 275–292 (1999)
37. Ha, U., Lee, L., Padmanabhan, V., Whang, S.: Manage. Sci. **43**(4), 546–558 (1997)
38. Beamon, B.M.: Int. J. Prod. Econ. **55**(3), 281–294 (1998)
39. Stevens, G.C.: Int. J. Phys. Distrib. Mater. Manage. **19**(8), 3–8 (1989)
40. Gunasekaran, A., Patel, C., Tirtiroglu, E.: Int. J. Oper. Prod. Manag. **211**(2), 71–87 (2001)
41. Sathish, T.: J. New Mater. Electrochem. Syst. **20**(4), 161–167 (2017)
42. Govindan, K., Soleimani, H., Kannan, D.: Eur. J. Oper. Res. **240**(3), 603–626 (2015)
43. Scholten, K., Schilder, S.: Supply chain management. Int. J. **20**(4), 471–484 (2015)
44. Liu, S., Chang, Y.-T.: Sustainability **9**(2), 222 (2017)
45. Thomas, D.J., Griffin, P.M.: Eur. J. Oper. Res. **94**(1), 1–15 (1996)
46. Seuring, S., Müller, M.: J. Cleaner Prod. **16**(15), 1699–1710 (2008)
47. Cooper, M.C., Lambert, D.M., Pagh, J.D.: Int. J. Logistics Manage. **8**(1), 1–14 (1997)
48. Cooper, M.C., Ellram, L.M.: Int. J. Logistics Manage. **4**(2), 13–24 (1993)
49. Croxton, K.L., Garcia-Dastugue, S.J., Lambert, D.M., Rogers, D.S.: Int. J. Logistics Manage. **12**(2), 13–36 (2001)
50. Srivastava, S.K.: Int. J. Manage. Rev. **9**(1), 53–80 (2007)
51. Carter, C.R., Rogers, D.S.: Int. J. Physical Distrib. Logistics Manage. **38**(5), 360–387 (2008)
52. Chen, I.J., Paulraj, A.: J. Oper. Manage. **22**(2), 119–150 (2004)
53. Keah Choon Tan: Eur. J. Purchasing Supply Manage. **7**(1), 39–48 (2001)
54. Teresa Teresa, M., Nickel, S., Saldanha-Da-Gama, F.: Eur. J. Oper. Res. **196**(2), 401–412 (2009)
55. Sarkis, J.: J. Cleaner Prod. **11**(4), 397–409 (2003)

56. Keah, C.T., Lyman, S.B., Wisner, J.D.: Int. J. Operations Production Manage. **22**(6), 614–631 (2002)
57. Ivanov, D.: Structural Dynamics and Resilience in Supply Chain Risk Management, pp. 293–313. Springer, Cham (2018)
58. Magee, L., Scerri, A., Cahill, F.: Reframing social sustainability reporting: towards an engaged approach. Environ. Dev. Sustain. 225–243 (2013)
59. Jiang, J., Stringer, J., Xu, X., Zheng, P.: A benchmarking part for evaluating and comparing support structures of additive manufacturing. In: 3rd International Conference on Progress in additive manufacturing, 196–202 (2018). https://doi.org/10.25341/D42G6H
60. Liu, J., To, A.C.: Deposition path planning-integrated structural topology optimization for 3D additive manufacturing subject to self-support constraint. Comput. Aided Des. 27–45 (2017). https://doi.org/10.1016/j.cad.2017.05.003
61. Jiang, J., Xu, X., Stringer, J.: A new support strategy for reducing waste in additive manufacturing. In: The 48th International Conference on Computers and Industrial Engineering, Auckland (2018)
62. Jiang, J., Xu, X., Stringer, J.: Optimization of process planning for reducing material waste in extrusion based additive manufacturing. Rob. Comput. Integr. Manuf. 317–325 (2019)
63. Sandhu, K., Singh, J.P., Singh, S.: Some investigations on the tensile strength of additively manufactured polylactic acid components. In: Advances in Materials Processing, pp 221–230. Springer, Singapore (2020)

# Chapter 10
# Supply Chain Management in the 3D Printing Industry as Exemplified by a Selected Organisation

Joanna Woźniak, Grzegorz Budzik, and Łukasz Przeszłowski

**Abstract** Simultaneously with the arrival of the Fourth Industrial Revolution, a particular attention started to be paid to introducing solutions based upon modern technologies in the line of production. Among them, there are, to mention, but one, rapid prototyping techniques, commonly referred to as well as 3D printing. The literature of the subject shows that, regardless of its numerous advantages, managers do not find it easy to accept this technology in production, which is caused by the absence of a clear model showing the business strategy most suitable for the purpose of implementing additive manufacturing, and also determining whether this strategy can be applied to all kinds of products. In connection with what is stated above, the analysis of the 3D printing industry worldwide is conducted in this chapter, and also the method of supply chain management, as exemplified by a selected organisation, taking under particular consideration processes exerting influence upon reducing the waste of raw materials, the loss of working hours and the misuse of financial means, is presented.

**Keywords** Supply chain management · Additive manufacturing · 3D printing · Production engineering

## 10.1 Introduction

In times of a strong market competition and dynamic changes in the market, introducing innovative solutions within an entire supply chain is becoming a general trend observed in economy. Purchasers order ever smaller batches of products, while, simultaneously, they place their orders ever more frequently. This

J. Woźniak (✉)
Faculty of Management, Rzeszow University of Technology, Rzeszów, Poland
e-mail: j.wozniak@prz.edu.pl

G. Budzik · Ł. Przeszłowski
Faculty of Mechanical Engineering and Aeronautics, Rzeszow University of Technology, Rzeszów, Poland

exerts influence upon the size of batches, and the fact they are becoming ever smaller is frequently a reason why it is being difficult to maintain production profitability. In connection with those changes, enterprises have started to pay a particular attention to optimising the costs of their activity by means of the application of modern solutions typical of the Industry 4.0 conception [1, 2].

Additive manufacturing technologies, also commonly known as 3D printing, are regarded in the industry as one of the most important production technologies of the period of a dozen or so recent years [3, 4]. All over the world, enterprises are investing their financial means in purchasing and developing computer programs, and also machines and devices applied in production processes and the quality control of products manufactured with the application of additive manufacturing technologies [5].

3D printing offers a number of advantages, and is a contemporary alternative to production as it hitherto has been (or a solution that complements it); in the latter one, a set of machines may produce a series of products and there is a direct connection between complexity and production costs. In turn, in additive processes, a set of elements may be produced in the form of a single item, which can reduce costs significantly. The more complex a part is, the more profitable it is to produce it with the application of additive methods is. Therefore, 3D printing is a highly promising industrial solution. First and foremost, it streamlines production upon request, which means that demand is better adjusted to supply. In addition to that, the application of 3D printing may reduce component warehousing costs. In previous research [6, 7] effort was made on to minimised the high-cost tool for machining with use of 3D printed tool to cut soft polymeric materials. Therefore, implementing additive manufacturing exerts influence upon an entire enterprise value chain [8, 9].

The literature of the subject shows that, regardless of its numerous advantages, managers do not find it easy to accept this technology in production, which is caused by the absence of a clear model showing the business strategy most suitable for the purpose of implementing additive manufacturing, and also determining whether this strategy can be applied to all kinds of products. Due to that, it frequently occurs that managers find it difficult to implement this technology in the production system of which they are in charge [10–12].

## 10.2    Analysis of the 3D Printing Industry

Analysing numerous reports relevant to additive manufacturing technologies, it is possible to ascertain that, in the recent years, the sale of 3D printers worldwide has been increasing rapidly. Terry Wohler's 2018 Report [5] informs, that in the year 2015, the number of devices sold worldwide amounted to 283,885 pieces, in the year 2016 to already as many as 424,185 pieces, and in the year 2017 to 528,952 pieces (refer Fig. 10.1).

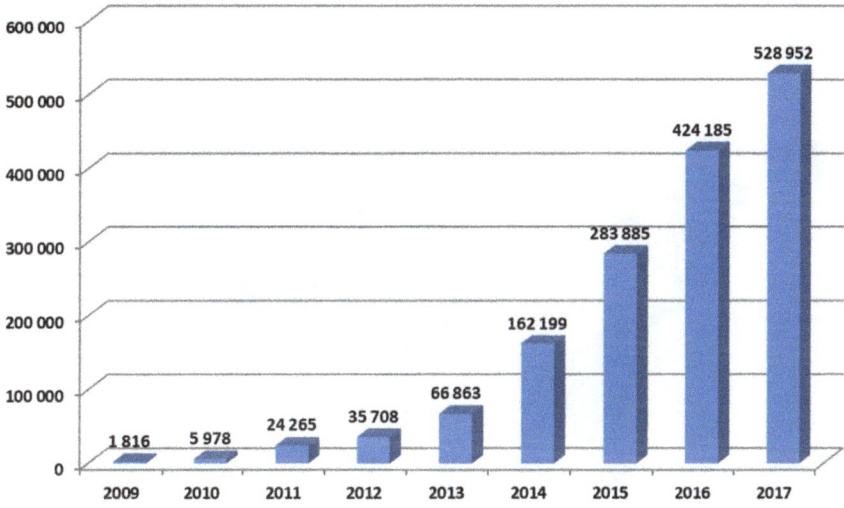

**Fig. 10.1**   Number of 3D printers sold worldwide in accordance with Wohlers Report [5]

In Fig. 10.2, the share (in per cent) of the industrial systems of additive manufacturing technologies installed in particular states since the year 1988 till the end
of the year 2017 is presented. The country boasting the highest value of this share
(i.e. the number of the implemented systems) is the USA (35.9%), which results
from the fact that the number of the producers of such machines is greater there than
anywhere else. The countries in the following positions were China (10.6%), Japan
(9.3%), Germany (8.4%), the UK (4.1%), Korea (3.7%), Italy (3.3%), and also
France (3.1%) [5].

As it is reported in Wohlers Report 2018, the greatest share in the market of
industrial 3D printers in the year 2017 was that of the Stratasys company (27.2%).
Analysing Wohlers Report 2015, one may notice a substantial fall in the share of
this company in the global market in the recent years (by 24.7%). In the year 2014,
the number of devices sold amounted to 6665, in turn, in the year 2017 the company sold approximately 4,100 industrial 3D printers. Analysing Fig. 10.3, and also
Fig. 10.4, one may as well observe that the 3D printers producers market is
becoming ever more fragmented each and every year and that there are ever more
producers of such machines [5, 13]. This fact results in a substantial increase in the
competition in the 3D printing industry.

In the year 2017, the principal areas of the application of 3D printing were: the
production of functional parts (33.1%), adjusting and assembling (16.9%), and also
educational and research work (10.8%). The results and share in per cent are
expressed in Fig. 10.5.

Upon the basis of the quoted research and the analysis of the literature, the
authors decided to describe the 3D printing position in product lifecycle more
precisely. Product life cycle is a notion within the realm of marketing theory, and it

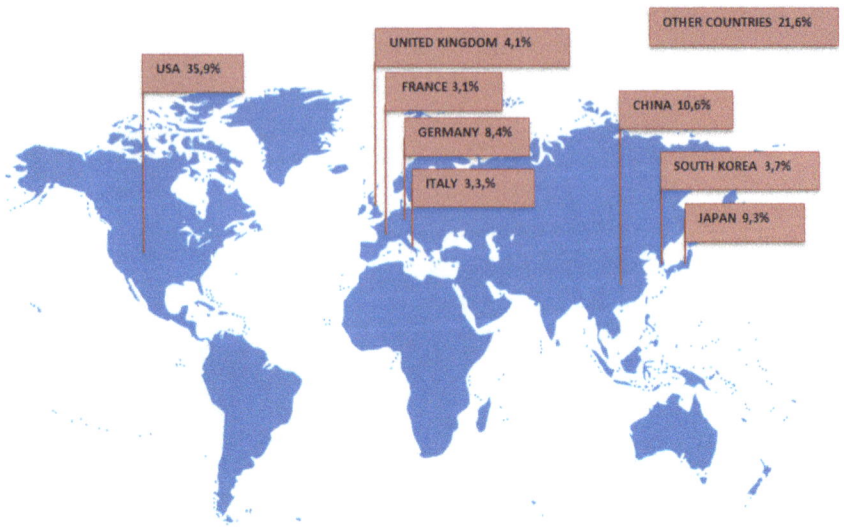

**Fig. 10.2** Share (in per cent) of particular countries in reference to possessed additive manufacturing technology machines [5]

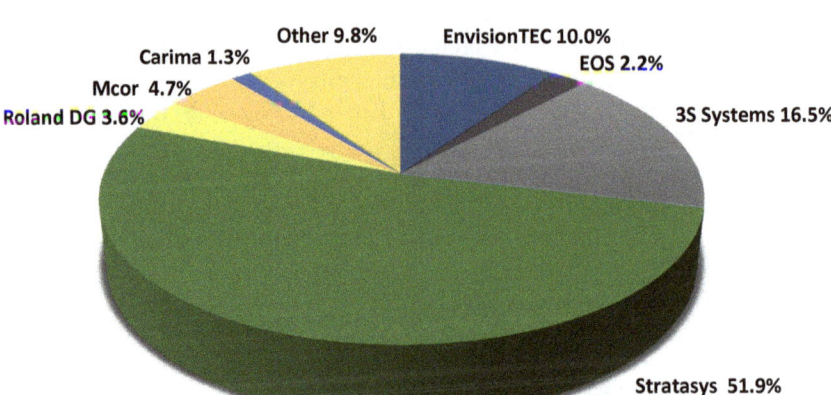

**Fig. 10.3** Share (in per cent) of the industrial producers of 3D printers—the year 2014 [13]

is most frequently composed of four phases: introduction, growth, maturity and decline (refer Fig. 10.6).

Figure 10.6 shows that the contemporary 3D printing industry is in the growth phase, whose features include a rapid pace of the application of streamlining solutions and innovations. The product is becoming ever more popular, and it is

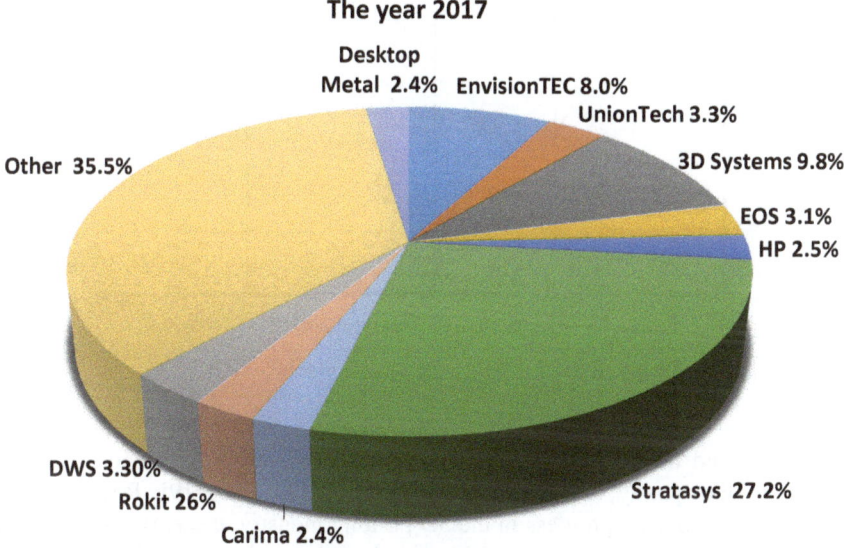

**The year 2017**

Desktop Metal 2.4%   EnvisionTEC 8.0%
UnionTech 3.3%
3D Systems 9.8%
EOS 3.1%
HP 2.5%
Other 35.5%
Stratasys 27.2%
DWS 3.30%
Rokit 26%
Carima 2.4%

**Fig. 10.4** Share (in per cent) of industrial 3D printers producers—the year 2017 [5]

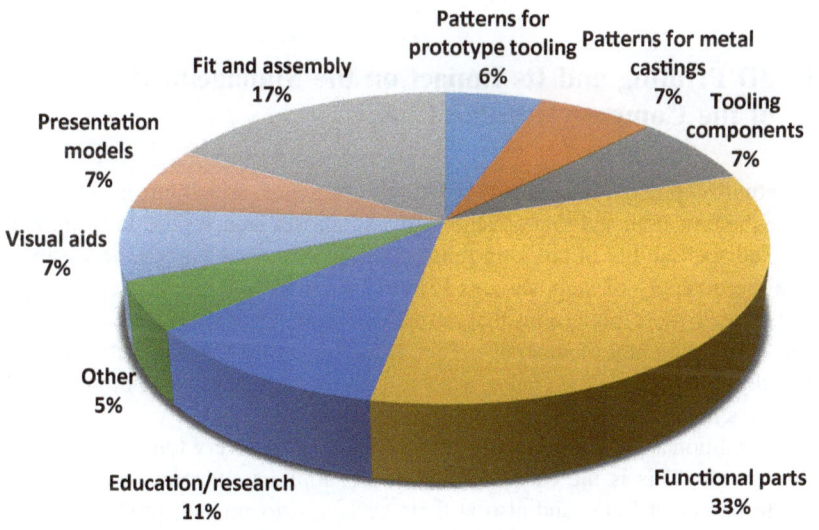

Patterns for prototype tooling 6%
Patterns for metal castings 7%
Fit and assembly 17%
Tooling components 7%
Presentation models 7%
Visual aids 7%
Other 5%
Education/research 11%
Functional parts 33%

**Fig. 10.5** The scope of the applications of additive methods and their share in per cent [5]

highly profitable to invest in it. The 3D printing technology keeps developing, there are ever more producers of the machines of this type, and interest in them is visible both in the industry and among natural persons. Changes are visible also as well in

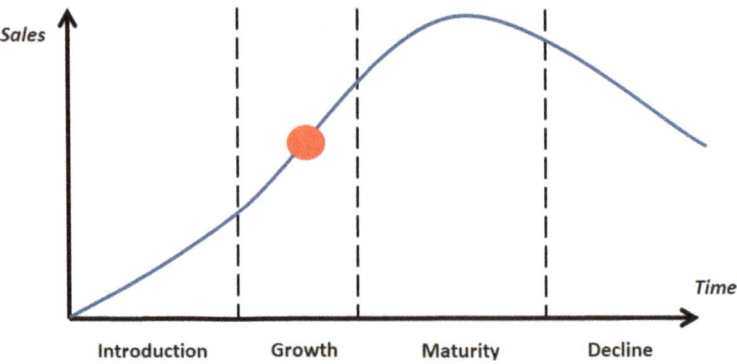

**Fig. 10.6** Position of 3D printing in product life cycle

the quality and diversity of printing materials, increasing productivity, and also improving the quality of manufactured elements [3, 4, 14–19]. Regardless of observing a significant progress in the 3D printing industry, it seems that it has not yet reached the phase of maturity (widespread implementing) [1]. Due to that, it is justifiable to monitor the market, and also to make streamlining changes within an entire supply chain.

## 10.3 3D Printing and Its Impact on the Management of the Company's Value Chain

Contemporarily, product life cycle is constantly shortened, which forces enterprises to reduce the time assigned to product design and market launch [20]. In addition to that, the indispensability of tailoring finished products means that it is necessary to increase the diversity of their variants [21–24]. Therefore, in production management, it is ever more noticeable that companies have to replace the economy of scale with the economy of diversity. Therefore, the organisation of vertically and horizontally integrated value chains plays a decisive role in terms of how fast and effective reactions to changes are.

In the traditional approach, supply chains are, as a rule, very long and made up of many levels. This is the case both as far as suppliers and producers are concerned. In particular links, and also at their '*rendez vous* points', product stock is maintained, which generates costs and renders it necessary to freeze enormous working capital [25–27]. As it is reported by K. Rutkowski and B. Ocicka, supply chain costs in production companies amount to, as a rule, 60–90% of all costs, and may even exceed 80% of revenues [12].

Therefore, the application of 3D printing may result in reducing and narrowing an external chain by means of reduction of the number of links involved in the process of developing product value. Thanks to that, it will become more simplified and clearer, and also, the borders between designing, production and distribution will lose some of their importance [8, 28].

The application of additive technologies in an enterprise may result, as well, in transition from centralised supply chains to the decentralised ones [29]. Production companies may organise purchasing indispensable raw materials directly, print a product and dispatch it to an end user. Then, developed the multi-layer structures of geographically dispersed suppliers (the companies of the processing industry and component suppliers) will disappear, consequently, reducing transactional and logistic costs.

## 10.4  Presentation of the Resources of the Rapid Prototyping Systems Laboratory of the Rzeszów University of Technology

The Rapid Prototyping Systems Laboratory, being a unit of the Department of Mechanical Engineering of the Rzeszów University of Technology, has been functioning since the year 2001. The cumulative collation gives rise to the conclusion that the Laboratory is in the possession of:

- 17 FFF 3D printers (15 Prusa i3 MK3S printers, 1 Ultimaker 3 Extended printer and 1 3D Gence INDUSTRY F340 printer),
- 4 MEM 3D printers (UP BOX+),
- 1 FDM 3D printer (Stratasys F170),
- 1 DLP 3D printer (Envisiontec Vida HD),
- 2 PolyJet 3D printers (Stratasys Objet, 3503D Objet Eden260V),
- 1 MSLA 3D printer (Peopoly MOAI),
- 1 DMLS/SLM 3D printer (Objet EOSINT M 270),
- 1 3DP printer (Z510 Spectrum),
- 1 SLA printer (SLA 250 3D Systems).

Among testing devices, there is as well milling machine (Haas Mini Mill 2), which is applied at the stage of post-processing.

In order to conduct the quality control of products/models manufactured additively, apart from popular measurement tools (such as a slide caliper, micrometre, and a toolroom microscope), a manual scanner (3D Creaform ACADEMIA 50), an optical scanner (ATOS Triple Scan II Blue Light of the GOM company), and also a coordinate measuring machine (Wenzel LH87), are applied. The equipment of the Laboratory is presented in Fig. 10.7.

It matters that the specific character of additive manufacturing renders it possible, in the case of 3D printers being parts of a production network, may be working on a single large order or a number of single-unit orders simultaneously [3]. Thanks to having such a large number of various machines, the Rapid Prototyping Systems Laboratory of the Rzeszów University of Technology may,

1) Prusa i3 MK3S

2) Ultimaker 3
Extended

3) 3DGence
INDUSTRY F340

4) UP BOX +

5) STRATASYS F170

6) Envision Vida HD

7) Printers PolyJet: Objet Eden260V and Stratasys Objet
350

8) Peopoly MOAI

9) Objet EOSINT M
270

10) Z510 Spectrum

11) SLA 250 3D
Systems

**Fig. 10.7** Equipment of the Rapid Prototyping Systems Laboratory of the Rzeszów University of Technology

12) Haas Mini Mill 2

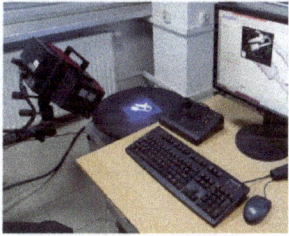

13) 3D Creaform
ACADEMIA 50

14) ATOS Triple Scan II Blue
Light of the GOM company

15) Wenzel LH87

**Fig. 10.7**   (continued)

therefore, simultaneously conduct research, scientific and instructional work, and also process orders for 3D printers from external customers. Production may be conducted not only on the scale of single units, but, as well, of series, which improves the chances in negotiations with customers because the deadline of processing an order may be set far before in comparison with that offered by competitors which cannot boast such a developed infrastructure. The Laboratory constantly broadens the scope of its activity by means of purchasing new 3D printers, devices for the 3D prints quality control, and also for testing reverse engineering materials and system. This fact exerts influence upon the possibility of conducting research work within the scope of additive technologies, and also allows to improve market position.

## 10.5   Presentation of a Logistics Network

In accordance with the literature of the subject while discussing the structure of a logistics network, one ought to take under consideration suppliers, the places of production, and also customers [26, 27]. Upon the basis of the profound analysis of the activity of the Laboratory, it was ascertained that the Rzeszów University of Technology applied the Pull system, in which activities connected with manufacturing process are commenced when an order is placed by a customer. This system

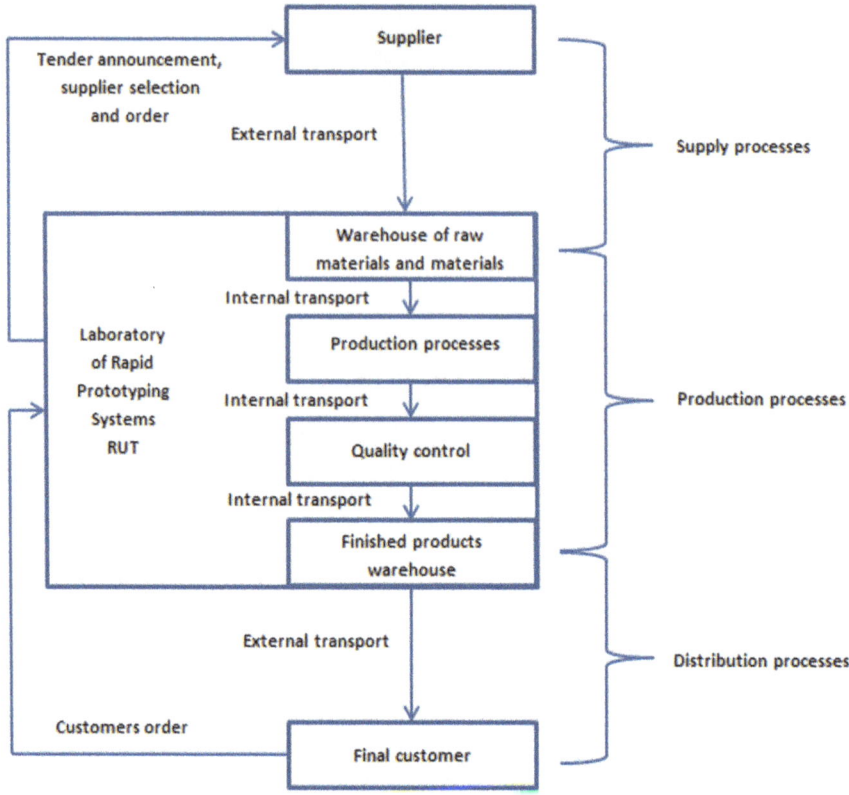

**Fig. 10.8** Presentation of a logistics network

is to 'make production process lean' by means of reduction of the costs connected, among others, with the excessive quantity of warehoused stock.

Figure 10.8 present the flow of raw materials and finished products typical of the Rapid Prototyping Systems Laboratory of the Rzeszów University of Technology.

Upon the basis of Fig. 10.8, it may be concluded that all activities connected with developing order plans, and also purchasing, are referred to as supply processes. All and any activities and operations in which a raw material is transformed into a finished product are production processes. In turn, distribution processes include the activities connected with the flow of goods from a finished products warehouse (with, or without, the participation of intermediaries) to a final customer.

Commencing the analysis of transport and manufacturing processes, a particular attention ought to be paid to enterprise's demand for raw materials. In accordance with the definition, needs in terms of material(s) constitute a direct demand for all and any materials resulting from planned production, and also auxiliary materials rendering it possible for an enterprise to function properly. Due to that, units responsible for planning and conducting purchase processes ought to possess detailed information about production plans. Therefore, information flow is one of the crucial elements of optimising a logistics network. In addition to that, placing large orders with a single supplier has a number of advantages. First and foremost, the purchaser is able to take advantage of the so-called economy of scale and improve their position in negotiations; thanks to a large order, a supplier may increase production, and, consequently, reduce the unit cost of produced materials.

After combining all needs in terms of materials, resulting from production plans, a tender is prepared and announced. In accordance with the Public Procurement Law [30], the principal criterion of selecting a company is the price of providing an ordered service. After collecting goods and conducting quality and quantity control, the goods may be placed in a warehouse or directly on a 3D printer holder (or in such a printer) in order to commence printing process.

In the case of a customer ordering that a 3D model be designed, rapid prototyping process is commenced by modelling the virtual geometry of an item in any CAD system. Next, a lattice geometry ought to be prepared, and that means, recording an item in a file, e.g. in the STL format. The next stage is that of setting the parameters of the printer, and also of dividing the geometry into layers (pre-processing). After these activities, programming the movements spreading a material, binder or a laser beam is commenced, to be followed by commencing process and manufacturing a real item (processing). The final stage of production is post-processing, which may be performed manually or with the application of devices (post-processing).

After manufacturing a real item, it is controlled in terms of quality. If it is found incompatible, printing must be commenced again. In the opposite situation, i.e. the expectations are met, it is located in a finished products warehouse or packed and prepared for dispatch immediately. If a number of pieces ordered by a single customer are bigger, it is usual practice to combine them into a single transport. The period required to prepare a dispatch may be extended, but the costs born by a customer are, as a rule, lower. In order to ensure professional delivery service, the Laboratory charges forwarding companies with this task.

It matters as well that while determining the deadline of processing an order, the Laboratory staff always take into account the time for possibly required error corrections or re-commencing production if an item manufactured does not meet customer's expectations. Thanks to that, a risk of losing orders, frequently caused by a failure to deliver on time, ipso facto breaking the contract, is reduced.

## 10.6 The Application of 3D Printing with Sustainable Development in Mind

In the long run, the application of additive manufacturing technologies is a step on the road towards new solutions, reducing the consumption of raw materials, and also improving the efficiency of the production sector [31]. The Rapid Prototyping

Systems Laboratory of the Rzeszów University of Technology constantly conducts in-house analyses of the 3D printing industry, and also R&D. Upon this basis, the authors differentiated between a few areas where the application of additive technologies may exert influence upon reducing the waste of raw materials, the loss of working hours and the misuse of financial means. The following were included into the group of the most important ones:

- Shortening the time of product design and market launch (time-to-market)—3D printing allows to manufacture prototypes in a short time and test them in real-life conditions. This way, engineers may make immediate changes in the design until the demanded result is achieved.
- Possibility of manufacturing a prototype/a finished product with the application of a single additive technology machine—a 3D printer may work without any other machine being applied, and does not exert influence upon processes involving other machines in production network. This fact is an additional asset for enterprises because it significantly reduces the phenomenon of bottlenecks, being one of the principal problems in traditional production, where they hinder increasing production efficiency.
- Possibility of the production of goods/prototypes/tools meeting the individual needs of a customer—by means of the application of 3D printing, it is possible to manufacture a tailored product, at a competitive price and in a short time.
- Eliminating the losses connected with spare parts warehousing—the application of 3D printing allows to manufacture a given element in a comparatively short time. Due to that, traditional warehouses may be replaced with the digital ones, in which file-based databases are kept to be applied when a customer notifies of their need.
- Reduction of waste in production process—3D printing technology is environmentally friendly, and, in comparison with traditional production, generates much less waste.

## 10.7 Proposals of Streamlining and the Directions of Further Research

The specification of the manufacturing process and manufacturing services as part of the Industry 4.0 concept makes the form of placing an order, the speed of the producer's response, and the order fulfilment time become a very important aspect for recipients [32]. An increasing number of companies strive to improve the ordering process of 3D models/products by introducing an Internet platform, where it is possible to pre-verify the 3D-CAD file and to conduct automatic valuation and initial simulation regarding the date of order fulfilment. The introduction of such a solution greatly simplifies the task for manufacturers, as well as increases the satisfaction of potential customers who receive the information they need almost immediately after sending the 3D-CAD file. An example is the treatstock.com platform, which operates on a global scale and brings together over 2,000 3D printing suppliers. Taking into account the customer's location and model requirements, it searches and compares prices from suppliers, and supports activities related to the automation of procurement processes [33].

A major problem in terms of increasing production efficiency (and, ipso facto, an entire supply chain) in additive manufacturing technologies is that the majority of 3D printers requires human labour input both before production process and after it [34, 35]. Due to that, production efficiency process with the application of additive manufacturing technologies depends, as well, upon the expertise, experience and predispositions of machine operators. Albeit there are already solutions which limit human labour input to minimum (such as Continuous Build 3D Demonstrator of the Stratasys company, which, after manufacturing a real item automatically moves a model to a mounted container, and, thanks to that, commences another printing process immediately) in the market, however, in spite of that, the majority of 3D printers available in the market require being operated [36]. Therefore, it is justifiable to design solutions the purpose of which is to minimise human labour input at each and every stage of the production of 3D models/products.

It ought to be borne in mind that the quality of 3D printing depends not only upon parameters of the printer itself and the operator's competences, but also upon the quality of a filament (printing material) [37]. For that reason as well, ever more frequently entrepreneurs wishing to control the quality of entire printing process and limit its costs decide to purchase or design their internal production line materials. Driven by concern for the environment, producers are working as well on solutions allowing to convert plastic waste into materials for 3D printing. Filaments are manufactured of plastic waste recovered from seabed, plastic bottles, and also old prints, and the remains of material after production process [38]. Filaments manufactured in 100% of plastics obtained by means of recycling meet all the requirements of sustainable development, and also reduce the impact of plastic on the environment surrounding us all [39, 40].

Due to the current epidemic threat, one of the global trends in production and resources management is, as well, the application of IT systems rendering it possible to work and monitor production process online. A 3D printer operator is obliged to monitor production process, thanks to which the risk of making an incompatible product is reduced (and so is that of damaging a machine). In the market, there are ready tools which allow to control and manage a 3D printer remotely from any place with access to the Internet. To mention, but one, the OctoPrint tool, which, together with a camera, allows to control and manage printing process with the application of a smartphone and in real time [41] can be referred to. In the Rapid Prototyping Systems Laboratory, research into the remote control of manufacturing process which may be implemented in the future, is also being conducted.

Therefore, in the context of the professional competences of engineers, it is becoming required that they possess interdisciplinary skills combining, To mention, but one, automatics, mechanics, robotics or IT. In addition to that, progress in internalisation renders it necessary for engineers to be acquainted not only with software dedicated to a given 3D printer, but, as well, to be able to communicate in foreign languages skillfully. These problems may be one of the critical points on the road towards increasing production efficiency in the context of the Industry 4.0 conception premises. For this reason as well, the employees of the Laboratory, by means of participating in numerous training events, conferences, and also industrial fairs, strive to develop their competences constantly, and also try to observe how competitors behave. In addition to that, the Rzeszów University of Technology is the organiser of the scientific conference 'Rapid Prototyping—Modelling—Manufacturing—Measuring in the 4.0 Industry Structure', which constitutes a major opportunity for industrial meetings in Poland. This conference provides an opportunity to exchange opinions and experiences relevant to the application of additive technologies, and also is a source of inspiration both for producers and customers from the 3D printing industry.

The authors think that the implementation of the recommendations described above may be an opportunity for enterprises to further develop and increase their competitiveness over other entities in the 3D printing industry.

## 10.8   Conclusion

Upon the basis of the completed tasks, the following conclusions were formulated:

- additive manufacturing technologies are a highly promising industrial solution, and, by means of the application of suitable processes and materials, they allow to manufacture fully valuable finished products;
- implementing additive manufacturing technologies definitely reduces the level of stock in production enterprises; therefore, it may be fully justified to apply the Just in Time system, and also (typical of it) Pull production. A supply chain based upon these systems allows a customer to ask whether a company produces what they need before production process is commenced;
- 3D printing brings desired results, especially if the properties of a product are dependent upon the customer's will to a great degree. Such types of production include: ETO—Engineering to Order, and also MTO—Make to Order. Therefore, it is possible to claim that contemporarily additive manufacturing technologies are not suitable for all cases;
- in comparison with traditional production, the application of 3D printing allows to manufacture a product much faster, consuming much less raw materials, a great deal fewer tools, much less space and human labour input, and also reduces waste level while simultaneously allows to re-process them, and, ipso facto, means that meet all the requirements of sustainable development;
- in order to increase effectiveness, it is recommendable to design solutions the purpose of which is to minimise human labour input in each and every link of a supply chain.

In the near future, the authors plan to continue research in the field of supply chain management in the 3D printing industry on a larger scale and make comparisons with other companies in the industry, which will allow for drawing deeper conclusions.

# References

1. Schwab, K.: The Fourth Industrial Revolution. Random House Lcc, US (2016)
2. Królczyk, G., Legutko, S., Królczyk, J., Tama, E.: Materials flow analysis in the production process—case study. Appl. Mech. Mater. **474**, 97–102 (2014)
3. Redwood, B., Schöffer, F., Garret, B.: The 3D Printing Handbook: Technologies, Design and Applications. 3D Hubs, Amsterdam, The Netherlands (2017)
4. Siemiński, P., Budzik, G.: Techniki przyrostowe. Druk 3D. Drukarki 3D. Oficyna Wydawnicza PW, Warszawa (2015)
5. Wohlers Report.: 3D Printing and Additive Manufacturing State of the Industry. Annual Worldwide Progress Report, Wholers Associates, USA (2018)
6. Sandhu, K., Singh, G., Singh, S., Kumar, R., Prakash, C., Ramakrishna, S., Królczyk, G., Pruncu, C.I.: Surface characteristics of machined polystyrene with 3D printed thermoplastic tool. Materials **13**(12), 2729 (2020)
7. Singh, S., Singh, G., Sandhu, K., Prakash, C., Singh, R.: Investigating the optimum parametric setting for MRR of expandable polystyrene machined with 3D printed end mill tool. Mater Today Proc (2020)
8. Mavri, M.: Redesigning a production chain based on 3D printing technology. Knowl. Process Manage **22**(3), 141–147 (2015)
9. Zhong, R.Y., et al.: (2017) Intelligent manufacturing in the context of industry 4.0: a review. Engineering **3**(5), 616–630
10. Fidali, M.: Szybki dostęp do części, czyli technologie przyrostowe w służbie utrzymania ruchu. Utrzymanie Ruchu **4**, 64–68 (2018)
11. Handal, R.: An implementation framework for additive manufacturing in supplychains. J. Oper. Supply Chain Manage **10**(2), 18–31 (2017)
12. Rutkowski, K., Ocicka, B.: Rozwój druku 3D i jego wpływ na zarządzanie łańcuchem dostaw. Gospodarka Materiałowa i Logistyka **12**, 2–11 (2017)
13. Wohlers Report (2015) 3D Printing and Additive Manufacturing State of the Industry. Annual Worldwide Progress Report, Wholers Associates, USA
14. Berman, B.: 3D printing: the new industrial revolution. Bus. Horiz. **55**(2), 155–162 (2012)
15. Budzik, G., Przeszłowski, Ł., Wieczorowski, M., Rzucidło, A., Gapiński, B., Krolczyk, G.: Analysis of 3D printing parameters of gears for hybrid manufacturing. AIP Conf. Proc. **1960**(1), 140005-1–140005-6 (2018)
16. Horst, D.J., Duvoisin, C.H.A., de Vieira, R.A.: Additive manufacturing at Industry 4.0: a review. Int. J. Eng. Tech. Res. **8**(8), 3–8 (2018)
17. Mellor, S., Hao, L., Zhang, D.: Additive manufacturing: a framework for implementation. Int. J. Prod. Econ. **149**, 194–201 (2014)
18. Ngo, D., Kashani, A., Imbalzano, G., Nguyen, K.T.Q., Hui, D.: Additive manufacturing (3D printing): a review of materials, methods, applications and challenges. Compos. Part B Eng. **143**(1), 172–196 (2018)
19. Wong, K.V., Hernandez, A.: A review of additive manufacturing. Int. Sch. Res. Netw. **4**, 1–10 (2012)
20. Kolberg, D., Zühlke, D.: Lean automation enabled by Industry 4.0 technologies. IFAC-PapersOnLine **48**(3), 1870–1875 (2015)

21. Chen, J., Reilly, R.R., Lynn, G.S.: Newproduct development speed: too much of a goodthing? J. Prod. Innov. Manage **29**(2), 288–303 (2012)
22. Hofmann, E., Rüsch, M.: Industry 4.0 and the current status as well as futureprospects on logistics. Comput. Ind. **89**, 23–34 (2017)
23. Wolniak, R.: The assessment of significance of benefits gained from the improvement of quality management systems in Polish organizations. Qual. Quant. **41**(1), 515–528 (2013)
24. Zimon, D., Malindžák, D.: Proposal of quality management and technology model supports a subsystem of manufacturing logistics. LogForum **13**(1), 19–27 (2017)
25. Antosz, K., Pacana, A., Stadnicka, D., Zielecki, W.: Lean manufacturing. Doskonalenie produkcji. Oficyna Wydawnicza Politechniki Rzeszowskiej, Rzeszów (2015)
26. Pagell, M., Shevchenko, A.: Why research in sustainable supply chain management should have no future. J. Supply Chain Manage **50**(1), 44–55 (2014)
27. Zimon, D., Madzík, P.: Standardized management systems and risk management in the supply chain. Int. J. Qual. Reliab. Manage **37**(2), 305–327 (2019)
28. Bogers, M., Hadair, R., Bilberg, A.: Additive manufacturing for consumer-centric business models: implications for supplychains in consumer goods manufacturing. Technol. Forecast Soc. Chang. **1**, 225–239 (2016)
29. Janssen, R., Blankers, I., Moolenburgh, E., Posthumus, B.: The impact of 3D printing on supply chain management. TNO **205**, 156–162 (2014)
30. Act of 28 January 2004—Public Procurement Law (Journal of Laws of 2018, item 1986, 2215)
31. Cunico, M.: 3D printers and additive manufacturing: the rise of industry 4.0. Concep3d (2019)
32. Siedlecka, S.: Porównanie i ocena jakości obsługi klienta dla wybranych firm kurierskich. Autobusy: technika, eksploatacja, systemy transportowe **12**, 1610–1613 (2017)
33. Treatstock. www.treatstock.com. Accessed 01 Oct 2020
34. Badiru, A.B., Valencia, V.V., Liu, D.: Additive manufacturing handbook book. Product Development for the Defense Industry. CRC Press (2017)
35. Bitonti, F.: 3D Printing design: additive manufacturing and the materials revolution. Bloomsbury Visual Arts (2019)
36. Oleksy, M., Budzik, G., Bolanowski, M., Paszkiewicz, A.: Industry 4.0 Part II. Conditions in the area of production technology and architecture of IT system in processing of polymer materials. Polimery **64**(5), 348–352 (2019)
37. Stratasys. www.stratasys.com/demonstrators. Accessed 01 Oct 2020
38. 3ders. www.3ders.org. Accessed 01 Oct 2020
39. Rebs, T., Brandenburg, M., Seuring, S.: System dynamics modeling for sustainable supply chain management: a literature review and systems thinking approach. J. Cleaner Prod. **208**, 1265–1280 (2019)
40. Zimon, D., Tyan, J., Sroufe, R.: Implementing sustainable supply chain management: reactive, cooperative, and dynamic models. Sustainability **11**, 7227 (2019)
41. Octoprint. www.octoprint.org. Accessed 01 Oct 2020

Lightning Source UK Ltd.
Milton Keynes UK
UKHW020604050922
408354UK00002B/44

9 783030 752378